True North

Also by Myron Arms

Servants of the Fish: A Portrait of Newfoundland after the Great Cod Collapse

Cathedral of the World: Sailing Notes for a Blue Planet

Riddle of the Ice: A Scientific Adventure into the Arctic

Touching the World: Adolescents, Adults, and Action Learning (with David Denman)

True North

Journeys into the
Great Northern Ocean

Myron Arms

 Upper Access, Inc., Book Publishers

Upper Access, Inc., Book Publishers
87 Upper Access Road, Hinesburg, VT 05461
(802)482-2988 • http:www.upperaccess.com

Cover design, interior design, and maps by
Kitty Werner, RSBPress, Waitsfield, Vermont
Cover photo by Jess Rice

ISBN: 978-0-942679-33-5

Library of Congress Cataloging-in-Publication Data

Arms, Myron.
    True north : journeys into the great Northern Ocean / by Myron Arms.
        p. cm.
    Includes bibliographical references.
    ISBN 978-0-942679-33-5 (alk. paper)
    1. Sailing--North Atlantic Ocean. 2. Sailing--North Atlantic Ocean. 3. North
Atlantic Ocean--Description and travel. 4. North Atlantic Ocean--Discovery and
exploration. I. Title.
    GV817.N73A76 2010
    797.12409163'1--dc22
     2009028346

Printed in the United States of America

∞ This paper meets the requirements of ANSI/NISO Z39.48-1992 (Permanence of Paper).

10   1 2 3 4 5 6 7 8 9 10

For John Griffiths,
whose skill and genius is everywhere apparent
in the able little craft that has so many times carried us safely
down to the sea and safely home again.

And for Amanda Lake, dear friend and shipmate,
in memorium

The end of all our traveling
Will be to return to the place we started
and know it for the first time.

T.S. Eliot

Northern Voyages of
Brendan's Isle

1984–86: Faroe Isles, Shetland, Scandinavia, Western Scotland, Southwestern Ireland

1987: Southwest Newfoundland, Gulf of Saint Lawrence

1988: Southern Labrador

1989: Circumnavigate Newfoundland

1991: Labrador

1994: Greenland

1998: Circumnavigate Newfoundland

2000–2003: Westmann Islands, Iceland, Faroe Isles, Shetland, Scandinavia

2004: Labrador

2007: Sable Island

2008: Labrador

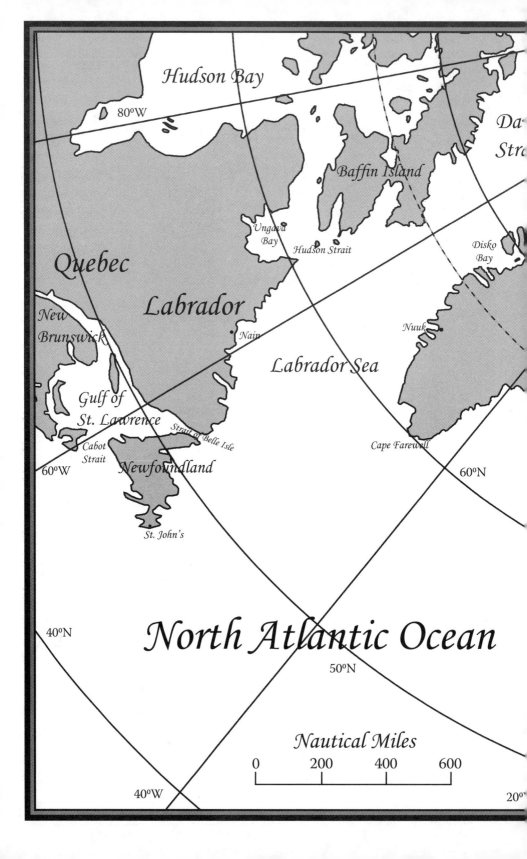

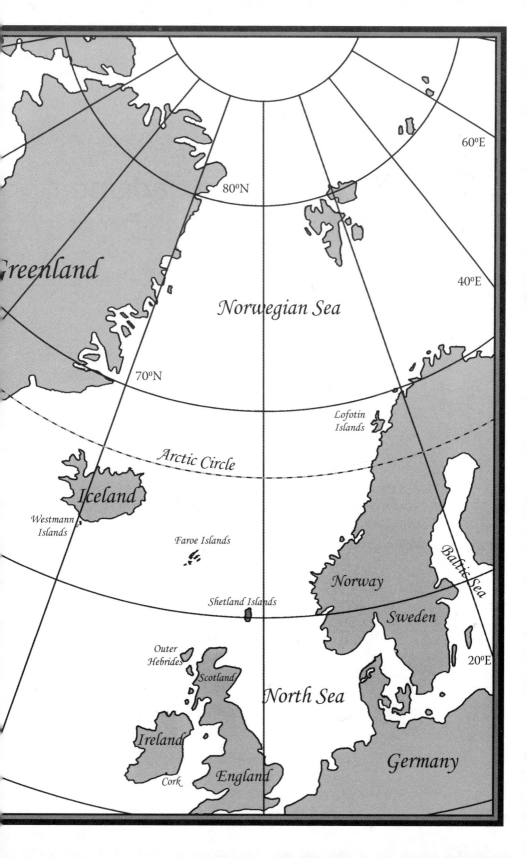

Text Acknowledgment

Certain of the essays in this collection first appeared in slightly different forms in the following publications, whose editors and staffs I wish to gratefully acknowledge: "Shakedown" in *Sail* (October 1984); portions of "Landfall Faroe Isles" under the title "Climbing the North Atlantic Hills" in *Sail* (February 1986); "Searching for the Edges" in *Sail* (June 1989); "The Wreck of the Braer" in *Cruising World* (June 1993); "Seeking Experienced Crew" in *Cruising World* (February 1995); portions of "Gale Warning" as the narrative format for Chapter 13: "Catalina," in *Servants of the Fish* (September 2004); "A Dune Adrift in the Atlantic" under the title "Skirting the Sands of Sable" in *Cruising World* (October 2008).

Contents

Introduction:
Perfect Travelers

I MAGINE A MOMENT IN THE Northern Ocean one morning fourteen centuries ago, in a time when the Earth had not yet been ravaged by humankind:

The surface of the water rises and falls in a lazy swell under a featureless gray sky. There is no sound of wind. The sun appears as a dim halo of light obscured behind ribbons of stratus cloud. Seabirds wheel and dive over a series of concentric rings in the water where a pair of whales has just sounded. Near the rings, shrouded in a veil of mist, moves a tiny ship, forty feet long and ten feet wide, fashioned out of animal skins and wood.

From a short distance away the ship looks like the discarded shoe of some gigantic walker of the Earth, covered as it is with rectangular slabs of blackened ox hide and sewn together with heavy thongs of horsehair, flax, and tallow. Brown flaxen sails are lashed to yardarms atop the two wooden masts, each sail painted with red ochre in the form of a Celtic cross. At the masthead above the larger of the sails, a long silken pennant hangs limp in the stagnant air.

A company of fifteen men lives in this ship. Several stand together this morning in the uncovered center section of the vessel, two facing forward on either side of the steering oar, four others facing aft and pulling together on a pair of long wooden sweeps. The remainder of the company rests in the ship's dank bowels where they endure the semi-darkness and the high animal odor of a place where too many men have lived and eaten and slept in uncomfortable proximity for too long.

The voyage these men have made is nearly six weeks old, although most of them have forgotten this fact by now. The continual parade of nights and days of wandering across ice-strewn seas has caused the measure of time to grow indistinct. Hours have passed into days and days have passed into weeks until there is no longer any certain means of counting the time. The men have sailed their ship to the accompaniment of rain and sleet and summer gales until it seems they will never see the land again... until this moment... this long awaited, terrifying moment.

Suddenly somewhere beyond the curtain of fog come the sounds of an unknown coast—the faint chatter of seals, the cry of gulls, the grinding of surf against a granite shore—sounds that stop the breath of every man aboard the ship. The dangers these sailors have faced for so long, dangers of storm and cold, hunger and fatigue, seem in this moment almost benign as they are replaced by the even more uncertain dangers of the land.

The leader of the company moves aft to take over the steering oar. He exhorts his shipmates to gather their resolve and to arm their spirits with hope. He utters a prayer of thanksgiving to the Christian God whom he is certain has brought them to this landfall, and he points the leather ship in the direction of the sound, urging his oarsmen onward.

The loom of a rockbound headland materializes slowly out of the mist as the helmsman maneuvers the leather ship into the mouth of a large bay. In another few minutes the air clears completely, revealing "a wide land, rich in fruit and flowers and autumnal trees." The helmsman propels the vessel deep into the bay until it is engulfed by the sweeping shoreline. Here he makes one final pull on the steering oar, driving the prow of the ship onto a beach of golden sand. He climbs to the gunwale, steps out onto the shore, and falls to his knees in prayer.

When he rises to his feet again, he surveys the shoreline and the lush forests beyond with a slow sweep of his eyes. Surely, he thinks, this is the land that has filled his imagination for so long—the land that Barrind described to him, the place Saint Mernoc visited so many times. Surely this is the green and fruitful glade that holy men and poets have for generations called the Promised Land, the "Blessed Isle of the Saints."

In this manner, on an anonymous summer morning in the sixth century of the Christian era, the Irish sailor-saint, Brendan of Kerry, is said to have achieved the object of his quest: landfall on the lush and elusive island referred to ever since as Saint Brendan's Isle. The tale of this voyage is recounted in the pages of a Latin text, written down several hundred years later and known throughout medieval Europe as the *Navigatio Sancti Brendani*. The story was widely popular during medieval times, attested to by the fact that more than a hundred hand-copied manuscripts still survive.

The tale continues in the words of the Latin text:

> *[Brendan and] the monks disembarked... [and] when they had gone in a circle around land, it was still light. They ate fruit and drank water, and in forty days' exploring did not come to the end of the land.*

The text concludes with reports of further exploration, with the discovery of a "great river" that divides the land, and with an encounter with a holy man who speaks to Brendan in his own language. This man cautions Brendan to return without delay to his home in western Ireland and foretells of the sailor's approaching death.

In a few more paragraphs the story ends. Brendan sails home and dies soon afterwards among his friends and disciples, but not before he is able to describe to them—and eventually to the world—all the wonders he has witnessed.

Ever since the writing down of the *Navigatio* more than a thousand years ago, the question has remained: Was Brendan's journey to the mysterious island a real voyage? And is the place that he is said to have discovered a real place? Archeologists and explorers, geographers and historians have argued about this question for generations. Cartographers during the first few centuries of the Age of European Discovery drew imaginary locations for Brendan's Isle in every corner of the North Atlantic Ocean. On some of the maps it was located where the Azores are now known to be. On others it appeared near Bermuda, Iceland, or Newfoundland. But on most of the early maps, it was simply drawn as a mysterious *terra incognita*, a land shrouded in fog somewhere out in the trackless waste of the northern ocean, surrounded by strange sea creatures and monsters of the deep.

In our own times there has been renewed interest in confirming (or debunking) the historical veracity of the voyages of Saint Brendan. In the late 1970s an intrepid young historian, Tim Severin, actually built a replica of Brendan's leather curragh and sailed it from County Kerry in western

Ireland to the Atlantic coast of Newfoundland in an attempt to prove the feasibility of such an undertaking. The story of his two-year odyssey was chronicled in a book, *The Brendan Voyage,* and later also appeared as a featured photo-essay in *National Geographic.* Yet even as Severin's modern voyage proved without a doubt that such a journey could be accomplished, it still stopped short of proving that a particular Irish cleric had made an actual historical voyage to a real place somewhere in the wilds of the western Atlantic.

In spite of all attempts to the contrary, it seems that the saga of Brendan and his fourteen "jolly saintes" belongs to the realm of legend—and here, in all probability, it will remain. Perhaps this is as it should be. Perhaps in the end it doesn't really matter whether there was an actual historical person named Brendan of Kerry who sailed an actual leather ship to the shores of the New World. The most important thing about this intrepid Irish explorer may simply be that we have his story. And his story, whether fact or legend, may be all we need.

The saga of Brendan's seven-year odyssey is unique in two important ways. Not only does it survive as the earliest written account of European contact with the unknown lands beyond the western ocean, but it also stands alone in the literature of European discovery as a testimony to a special kind of journey—a journey of the imagination and the spirit. Brendan and his crew set out in search of the Blessed Isle without economic motives or self-aggrandizing schemes. Their purpose was not to lay claim to new territories, subjugate native populations, or seek for gold or riches. They weren't looking for a northwest passage to the Orient. They weren't planning to trap for furs or establish a summer fishing station. They had, in fact, no other purpose for making their voyage than to witness for themselves the grand and terrible beauty of a wilderness they had heard extolled in legend but had otherwise only vaguely dreamed.

"The themes [of the *Navigatio*]," writes American nature writer Barry Lopez, "are of compassion, wonder, and respect," as opposed to the more familiar themes of bloodshed, plunder, greed, and conquest that typify so much of the history of early European contact with the American continent. Nowhere else in the long chronicle of European discovery is there a tale of another voyage quite like this one.

The Icelandic Sagas of the tenth and eleventh centuries describe several attempts by early Norse explorers to wrest fertile lands from their aboriginal inhabitants and to establish permanent trading colonies of their own on the shores of the New World. Four centuries later, the record of the

Columbus voyages echoes many of the same themes and establishes the rule for all who would follow—Cortez and Balboa, Verazanno and Cabot, Frobisher and Cartier—as they set out to plunder the American continent or to discover lucrative trade routes to the wealthy markets of the Far East. Hundreds (eventually thousands) followed in the footsteps of these early explorers, virtually all of them seeking land or power or economic gain.

The *Navigatio* of Saint Brendan, in contrast, is the tale of a voyage with a simpler objective. As Lopez again observes, it is the story of a group of "impeccable, generous, innocent, attentive men [who] were, one must think, the perfect travelers." Their journey was a pilgrimage of sorts, yet they did not sail as proselytizers or missionaries. Instead, they pursued their quest as celebrators of the Earth and as chroniclers of its awesome majesty. It was enough for these travelers to bring home memories of what they had seen and afterwards to tell the story of their voyage to all who would listen.

I first learned about the voyages of Saint Brendan more than two decades ago, during a time when I was building a sailboat in my back yard. With my wife Kay and our friend John Griffiths, I worked full time on that boat for almost a year. In the evenings I would read a few pages of Severin's *The Brendan Voyage*—that is, when I could stay awake long enough to read anything at all. Then, when daylight came again, I would head back down to the boat for another ten or twelve or fourteen hours of work.

For much of that year the sailboat had no name, although I was too busy at the time to worry about such secondary details. The name would come, I knew, in its own good time. Meanwhile there were more important things to worry about: things such as designing plywood templates for bunks and settees, ordering the next three thousand board feet of Philippine mahogany, cutting out seven hundred cypress plugs on the drill press, gluing up thirty-five cupboard doors, building the icebox and the navigator's table and the companionway stairs.

I knew what I was hoping to do with this boat once she was finished and launched. I wanted to sail her to the lonely coasts around the rim of the North Atlantic basin. I wanted to explore some of the natural places and see some of the wonders that old Saint Brendan himself may have witnessed during his seven years of wandering about the northern ocean. It therefore came as no surprise when the boat's name suddenly occurred to me—not as an idea but as an incontrovertible fact, as something that had always been. It came while I was working one morning fitting a piece of cabinetry in the main saloon. I was staring at the plank in my hand, concentrating on the asymmetrical sweep of its grain, preparing to mark with a pencil and compass the line that it would take as it fit against the irregular inner surface of the hull. One moment there was nothing in my mind but the task before me. The next moment there was a name: *Brendan's Isle.*

I felt as if I'd just emerged from some kind of temporary amnesia. Surely, I thought, I'd seen this name before. Maybe it had been written on the bill of lading from the shipping company that had delivered the empty hull and deck from the shipyard in Taiwan. Maybe it had been stamped onto the pine crates containing the spars and standing rigging that had come from New Zealand. Or maybe it had always been there, printed on the boat's transom in some kind of invisible braille, and I had only just now learned how to decipher the shape of the letters.

Brendan's Isle: the name had a romantic ring to it, a suggestion of faraway lands and olden times. It brought to mind the geography of the northern Atlantic—an apt connotation in light of the plans that were already forming in my mind about the places I wanted to sail. The name also suggested good luck. Saint Brendan, by all reports, had been an unusually lucky sailor. His chroniclers often used the phrase "Brendan luck" when describing the knack he seemed to have for avoiding dangers, weathering storms, surviving crises, finding safe landfalls.

On the more practical side, the name *Brendan's Isle* was short and unambiguous. On a radio call it would be quick to say. On a customs declaration it would be easy to spell. People would be able to remember it—unlike so many exotic yacht names that are virtually impossible to read or pronounce.

In short, *Brendan's Isle* had all the necessary attributes a sailor seeks when he wishes to name a little ship. The phrase was both beautiful and simple. And it seemed to speak to the particular purpose of the boat I was building. It was a name for the grand and mysterious places my shipmates and I would one day be setting out to find, the questions we would be trying to articulate, perhaps even some of the answers we would be hoping to find.

By July 1983 the building project was far enough along that we were able to launch and sail *Brendan's Isle* for a month, from the Chesapeake Bay to the southern coast of New England and home again, in order to sea test her systems and shake down her rig. The following winter Kay and I worked (without John) for another fifteen hundred hours, completing all the final

Details of the construction of Brendan's Isle's *main cabin*

modifications that the boat would need in order to be ready to take on the North Atlantic Ocean the following spring.

Brendan's sailing plan for the next two years began with a high-latitude crossing of the North Atlantic, a route that took her out past the Grand Banks of Newfoundland, across "iceberg alley," close past the southern capes of Greenland and Iceland, and on toward the coasts of northwestern Europe. Her first landfall, some twenty-three days and three thousand nautical miles after leaving the Chesapeake Bay, was in the Faroe Isles, a mountainous archipelago several hundred miles east of Iceland and almost the same distance due north of Scotland. The Faroes were a convenient stopping place for *Brendan's Isle* and her crew on this frigid west-to-east crossing of the North Atlantic, just as they may also have been for old Saint Brendan as he made his way out into the wilds of the western ocean while sailing in the opposite direction. If, as Tim Severin argues, the old sailor-saint followed the so-called "stepping stone" route across the North Atlantic, he would almost certainly have visited the Faroes (preceded by landfalls in the western Hebrides of Scotland, and followed by visits to Iceland, southern Greenland, and perhaps also the Atlantic coasts of Labrador or Newfoundland). According to the *Navigatio*, Brendan and his shipmates made landfalls early in their travels in a pair of hauntingly beautiful places described variously as the "Isle of Sheep" and the "Paradise of Birds." Both

The ruins of an ancient Christian chapel at Brandarsvik, Faroe Isles

names strongly suggest the Faroe Isles—as they surely must have appeared a millennium and a half ago and as they still appear today.

I had no conscious intention, as my shipmates and I started off across the Atlantic in the spring of 1984, of following in the footsteps of the old Irish sailor-saint. The legend of Brendan's voyage was engaging—the idea of a mysterious island somewhere out in the middle of the Atlantic was romantic and exciting. These were reason enough, I felt, for the name that appeared on the transom of our little sailboat. I had no thought as our first summer's sailing plans evolved of following specific routes that Brendan may have followed or of visiting particular places that he may also have visited.

Yet the morning that *Brendan's Isle* sailed up into Hestur Sound, Faroe Isles, and anchored near a place called Brandarsvik ("Brendan's Creek"), I admit I felt a strange sense of oughtness, as if we had sailed all these three thousand miles just to be in this setting. There was an aura about this place, a ghostlike presence that seemed to bridge the centuries and transform legend into living fact. The ruins of an ancient Christian chapel hunched on the beach, casting the shadows of its crumbling stone revetments on the sand. The dark green hills behind the ruins were dotted with sheep. A thousand feet overhead a thousand seabirds circled in silence, and for a moment I felt as if I could see the silhouette of a tiny leather ship lying at anchor near the mouth of the creek and the spectral forms of fifteen ragged sailors huddled at the chapel gates, seeking advice from the holy man within about where next to point their bows in their quest for the Blessed Isle.

Two and a half decades have passed since that day in Brandarsvik—two and a half decades during which I and *Brendan's Isle* have voyaged over one hundred thousand sea miles and have visited, in utterly haphazard fashion, virtually every one of the "stepping stones" that the old Irish sailor-saint might arguably have visited during his (equally haphazard) meanderings about the northern ocean. According to the *Navigatio*, Saint Brendan pursued his quest for seven years, ranging about the Atlantic each summer in a primitive and ungainly vessel whose speed and course were often dictated by the vagaries of wind and weather, and whose ultimate destinations were therefore often left entirely to chance. Each year at the end of their summer's voyaging, he and his shipmates would seek safe haven to rest themselves and to repair and reprovision their little vessel. Then, the following spring, they would set off once again with a renewed sense of purpose and an ever stronger hope that somehow, this time, they might happen upon the object of their quest.

In contrast, our travels aboard a modern, well-found yacht, efficient in its sail plan and equipped with all the miracles of modern electronic navigation, have been much more predictable. Each of the forays I've made into the North Atlantic during the past twenty-five years has been carefully conceived and painstakingly planned. Each has had a cruising agenda; each has had a timetable; each has had a set of predictable destinations. Twice I, with a small crew, have taken *Brendan's Isle* from the eastern seaboard of the United States across the top of the Atlantic to Scandinavia, both times calling at the Faroe Isles (and both times visiting the ruins at Brandarsvik).

On the first of these journeys, after a winter layover in southern Denmark, we crossed the North Sea to spend a month exploring the western coast of Scotland, and another month visiting the bays and sounds of southwestern Ireland. On the second, we paused to visit the Westmann Islands, south of Iceland, then to circumnavigate Iceland in its entirety. And both between and after these journeys, we visited the Atlantic coasts of Newfoundland and Labrador on numerous occasions, one of which included a mid-summer crossing of the Labrador Sea and an exploration of western Greenland all the way to the great ice fields of Disko Bay, three hundred miles north of the Arctic Circle.

A view of Brandarsvik from the road above the settlement

In the beginning, as I've said, I had no conscious intention of following in the footsteps of the old Irish navigator, retracing his voyages, visiting all the strange and marvelous places he might have visited. But fifteen years later, as I began to reflect on the record of my own travels, I realized that *Brendan's Isle* and I had in fact been pursuing an unspoken pilgrimage of sorts, stringing together a series of voyages that may have seemed random and disconnected when considered separately, but that began to form a compelling pattern when considered as a whole.

Eventually, by the summer of 1998, I realized there were only a handful of stopping places along the old "stepping stone" route that *Brendan's Isle* had not visited, and I set about to rectify the oversight. That summer, during our second circumnavigation of Newfoundland, Kay and I and our shipmates visited the tiny village of Saint Brendan (as good a candidate as any for the actual historical site of Brendan's New World landfall). Two years later, during our second eastbound trans-Atlantic crossing, we paused to call at the volcanic Westmann Islands, south of Iceland, there to witness the eerie, desert-like nakedness of Surtsey, a recently formed volcanic pinnacle that may be a modern counterpart to Saint Brendan's "Island of Smiths," the smoke-belching island in the *Navigatio* whose inhabitants hurtled hot, molten rocks at the monks as they sailed past.

Finally, two decades and more after we had first made landfall at the ruins at Brandarsvik, I began to feel as if I and all those who had sailed with me were on the cusp of completing a quest that we had been pursuing, consciously or not, for many years. We had sailed for one hundred thousand sea miles around the rim of the northern Atlantic, and now, with our circumnavigation of Iceland, we had managed to assemble a sequence of landfalls that might easily have been made by our legendary predecessor—and thus to have fashioned a modern "Brendan voyage" of our own.

The essays that follow are intended as a chronicle of this voyage—or really this series of voyages—and as a celebration of the natural places that *Brendan's Isle* and her crews and I have visited along the way. They are offered, with all humility, in the spirit of those perfect travelers, those "impeccable, generous, innocent, attentive men" who were satisfied simply to experience for themselves all the wonders of the natural world and afterwards to tell the story of their voyage.

There will be no overt political agenda here—with one important caveat. The "pristine wilderness" that Saint Brendan and his shipmates discovered in their travels and that were later described in the pages of the *Navigatio* no longer exists. Much as we might wish it otherwise, we, as a species, have succeeded beyond our wildest expectations over the past thousand years in taming the wild places, subjugating nature, establishing our imprint upon virtually every square inch of this planet. Some of this imprinting may be relatively harmless and inconsequential—styrofoam coffee cups washing up on the beaches of Antarctica, broken soda-pop bottles disintegrating into sea glass on the ocean floor. But much of it—the vast preponderance of it—is far more onerous. As we, the seven billion, have succeeded in taming the wilderness, we have also begun to alter it, often to the detriment of both the other species who live there and of ourselves.

Of necessity, the essays that follow are therefore set within an inevitable if often unspoken context of expanding human waste-streams, changing atmospheric chemistry, disappearing species, rising sea surface temperatures, thinning sea ice, melting glaciers. These things are. They exist as part of the landscape now—so that to claim to bear witness to the beauty of the natural places without also recognizing these threats to their survival is to fabricate a lie. Much as Rachael Carson did more than fifty years ago, I have also learned that it is no longer possible to celebrate the intricate natural rhythms of "the sea around us" without also becoming aware of the human-induced changes that threaten those rhythms from every side.

The age of innocence is past. We can no longer roam the Earth as old Saint Brendan may once have done, marveling in childlike wonderment at a pristine and unaltered nature. Instead, if we are honest with ourselves, we now must travel as my shipmates and I have tried to do aboard *Brendan's Isle*: aware of the changes that are taking place, deeply appreciative of the beauty that remains, armed with a kind of urgency that moves one at every moment to encounter the natural world, to live deliberately within it, to strive to minimize one's footprint upon it, and to bear witness to it before it is altered irretrievably—before it is lost.

Part I

Setting Out for Ithaka

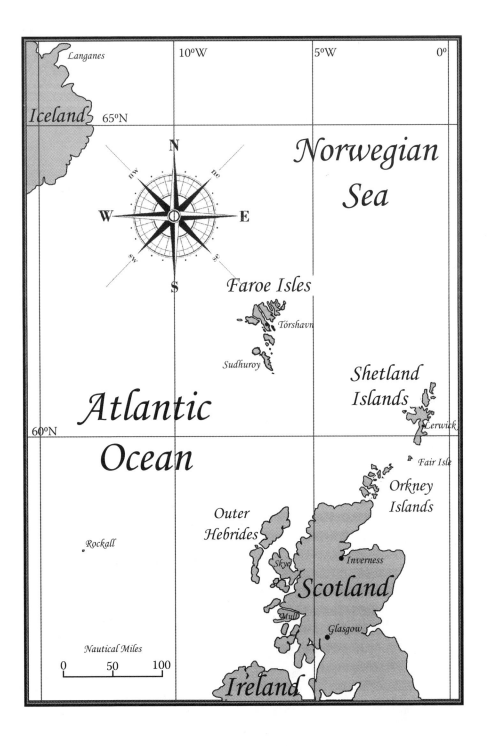

Every long ocean passage, before all else, is an act of imagination. The following essay was written in the dead of winter, six months before Brendan's Isle *was to begin her first north Atlantic crossing, while the enterprise was still only a vague collection of expectations and desires in the minds of those who would sail.*

1 Shakedown

OVERHEAD THE SKY IS BLACK. The stars shiver in the cold. The air temperature on deck skids down through freezing, with crystals of ice forming where there was moisture just moments before. I hang a kerosene lamp on the headstay of *Brendan's Isle* and check the anchor before heading below.

In the saloon the diesel heater begins to draw. I join four others around the dinette table, and as the cabin temperature rises, we begin a ludicrous striptease to the rumble of the fire. Layer by layer the outer wear comes off—parkas, heavy oilies, caps, mittens, sweaters, wool trousers, boots, wool boot liners, insulated shirts—all thrown in tumbled piles on the port settee. I am comfortable at last in cotton longjohns and a turtleneck.

The teapot whistles on the galley stove. John twists the cap off a new bottle of Mount Gay rum. He arranges a row of tea mugs on the table and fills them with hot water and rum. He raises his mug in a toast to the crew. Then a toast to the vessel—she sailed fast and dry today. And finally, a toast to the Faroe Isles—may there always be a gentle lee and safe harbor in their fiords.

It is easy to imagine *Brendan's Isle* at anchor this very night in one of those fiords of the Faroe Isles, perhaps on the island of Streymoy, under the lee of black granite hills. From somewhere outside come the cries of curlews and Arctic terns from their nesting places on the cliffs above. A river sparkling like diamonds tumbles down the rock past phalanxes of sheep grazing in the valley. A pale Arctic sun hovers over the jagged terrain. These are the islands I have dreamed of visiting under sail, islands that the fabled sixth century Irish navigator, Brendan of Kerry, may have visited and named the "Paradise of Birds," islands stark and pristine, that have come to represent for me a remote and irresistibly beautiful wilderness.

These are the islands where our vessel *will* be anchored, if all goes well, in another six months' time. But tonight is still winter half a world away, and *Brendan's Isle* lies at anchor near a creek mouth in the Sassafras River, Maryland, after a brisk shakedown sail in the upper Chesapeake Bay. The five who are aboard tonight are the same five who plan to sail the passage to the Faroes (and then, by way of the Shetland Islands, to Scandinavia) next summer. As skipper, I have organized this mid-December crew meeting both to tune the boat and to tune our minds and spirits for the challenges ahead.

As the first and longest leg of a high latitude north Atlantic crossing, the passage to the Faroe Isles promises to be an exciting ride. These sub-Arctic islands lie in latitude sixty-two degrees north, about two hundred fifty miles shy of the Arctic Circle. The most direct offshore route from the northeast coast of the United States to the Faroes will take a small sailboat out past Newfoundland to the Grand Banks, and then, through the certainties of ice, fog, and fast-moving North Atlantic depressions, along a great circle that will curve several degrees southeast of Reykjavik, Iceland, before homing in on Streymoy and the other seventeen islands of the Faroese archipelago.

Trees do not grow in the Faroes. The weather there in June is often damp, windy, and raw. A parade of cold westerly gales and green sub-Arctic seas will chase *Brendan's Isle* across many of the three thousand sea miles she will need to sail before landfall. The passage promises to be a thorough test of crew, vessel, and gear. The vessel and gear are nearly ready. This evening I want to be sure that each member of the crew is ready as well.

The conversation on deck all day has focused on practical matters—where to lead safety lines for the harness tethers, where to sheet the trysail, how to rig warps astern, how and in what sequence to change down the head rig, where to lash baggywrinkle on the backstays, where to sew chafe patches on the mainsail. Around the saloon table the same kinds of practical concerns dominate the conversation at first. We talk through several options for watch rotation and cooking responsibilities. John, who is the veteran of the group with three previous Atlantic crossings under sail, asks about medical experience. Who can stitch up a gash wound? Set a bone? Administer a needle? Who has taken recent instruction in CPR? Who can serve as dentist?

First a toast to the vessel…she sailed fast and dry today.

Kell, my oldest sailing pal and one who has looked forward to this journey almost as long as I have, asks about various responsibilities for maintaining machinery and equipment. Who will be engineer for the diesel? Who will be bosun? Sailmaker? Plumber and electrician? Kell volunteers to bring his scuba gear and be responsible for overboard repairs. I know from other passages we have sailed together that he will also volunteer to be the "monkey man" who goes aloft in the bosun's chair for any needed repairs to the rig while we are at sea.

Andrew, at twenty-two years old, is the second youngest on the crew. He will graduate from the University of Massachusetts this coming spring, and, along with all his formal education in the past few years, he has also managed to gather many thousands of miles of cruising, racing, and offshore delivery experience. He pours himself another rum and hot water and asks a question that has obviously been weighing on his mind. How about food? Who will plan menus? Who will stow the various items so we can find them when we need to?

Everyone laughs, but Andrew remains stone-faced. Food, his expression warns us, is no laughing matter. We need a "steward," he suggests— someone who will monitor menus from week to week and keep the galley stocked from stores hidden in the remote ends of the vessel.

Steve has been unusually quiet all evening, but suddenly he chimes in with a loud second to Andrew's suggestion. Steve is my youngest son and, at sixteen years old, the youngest member of this crew. He is a fine sailor with more experience than most who are many times his age, but I sense that he has been a bit overwhelmed by all the technical conversation. He is normally talkative and relaxed; today he has been preoccupied with silent questions about his own qualifications and mental readiness for the challenges ahead. When he volunteers to be Andrew's "steward," he seems to speak as a novice among seasoned journeymen, deferring to those with more experience in jobs that require specialized skills.

For several months I have been thinking about Steve's role in this undertaking. I want somehow to be sure that if he sails this passage, it will not be just because I have asked him and obviously want him along but because he wants the experience for himself. I remember vivid moments at sea with him when he was a small boy—golden sunrises, sudden encounters with whales and porpoises, long dark nights when he would not leave the helm until the end of his trick. I want to be sure that next summer's north Atlantic marathon is the proper next step for Steve—that it is in sequence, not too much too soon. I want him to be able to grow, as I suspect each of the others on this crew has grown, into an endless desire for seagoing encounter.

The rumble of the diesel fire and the warm rum soon combine to work their magic on *Brendan's Isle's* five inhabitants. I set my pad and pencil aside and snap off the bright overhead bulb. Kell lights the oil lamps on either side of the dinette table while John and Andrew begin preparations for dinner. Warm, safe at anchor, in the company of shipmates and friends, each feels the ancient atavism, the echo of some preconscious memory, perhaps, of safe haven around a communal fire in the caverns of pre-history. Unhurried, apart from the rest of the world, these are the times for inner searching and reflection.

"All of you should know that Steve and I have been talking pretty seriously about this journey," I say. "About a month ago, I asked him to reconsider—to search his own motives and decide whether he really wants to do this crossing—for his own reasons, not for anyone else's.

"The same goes for all of us. If you have any second thoughts, now is the time to say so."

I glance over at Andrew. "No second thoughts," he says simply.

John speaks up. "Maybe we should think about going even farther north," he says with a grin that reveals several of the broken teeth that he

The author at the navigation station plotting the day's course

received thirty years ago in a near-fatal fall in the winter Alps. "I mean north around Iceland—the Denmark Strait is usually free of pack ice by late June. If we push north of the circle, we'll see the midnight sun at the solstice. That would be something worth seeing, lads!"

Does John really mean what he is saying? I try to read his expression as he stirs a bag of instant rice into a pot on the stove. I know John well—not just as a friend, but also, more recently, as a teacher. John is a talented shipwright and cabinet-maker. We have just finished twelve intensive months working side by side, as master and apprentice, building the accommodations of *Brendan's Isle* from a bare hull and deck. That project was something of a marathon in itself, involving more than thirty-five hundred hours of work between us. We talked often and long during those months, and I think I know some of the ghosts and some of the dreams that feed his imagination.

Once upon a time, back in the years when John used to divide his time between yacht deliveries and alpine climbing, he was asked by the legendary Bill Tillman to sail aboard *Mischief* to Greenland and Baffin Island in the Davis Strait. Tillman's timetable was sufficiently vague that John was away on a yacht delivery when *Mischief* set off, and he missed an adventure of a lifetime—or so he felt. Now comes another chance to sail north—to see the glow of the northern sun at midnight, to sail with minke whales and bluewhite bergs, to hear the shriek of northern gales in the rigging, to climb into the sub-Arctic hilltops of the Faroes and breathe an atmosphere as frigid and pure as any on this planet.

I know there is no way we can afford the time to detour five hundred miles north around Iceland and also be in the southern Baltic in time for a summer cruise, but I realize John is dead serious when he talks of Iceland and the Denmark Strait.

"I wish we could, John. I wish we had the time."

"Maybe you ought to read that poem of yours again," says John. He refers to a paraphrase of the C.F. Cavafy poem "Ithaka" that I mailed to each of this group in my last crew letter. A copy of this poem is taped right now to the bulkhead near the companionway steps, where it will be framed and hung permanently for the crossing. I ask John to read it aloud if he'd like, as he is standing closest to the stairway. He pauses—then reads the words:

> *As you set out for Ithaka*
> *Ask that your journey be long and hard and full of adventure.*
> *Along the way keep Ithaka always in your mind,*
> *But do not hurry the journey at all—*
> *Better if it lasts for many years.*
> *When you arrive, wise and full of experience,*
> *You'll find her poor, with not much left to give you.*
> *No, she'll have nothing to give you,*
> *But she will have given the voyage,*
> *And that is something; perhaps that is enough.*

"Am I a shameless hypocrite then?" I ask the question out loud. Maybe I am, for I find myself dreaming continually of the Faroes these days. Every time the telephone begins its incessant ringing, every time I glance at another headline about chemical waste or acid rain or a new oil spill somewhere in the world, I find myself focusing on that little cluster of hashmarks on the chart of the North Atlantic as if it were some kind of magical Shangri-La. "Maybe it's just that I am not yet wise enough and 'full of experience,'" I say to John. "Maybe I'll need the voyage itself for that."

"I don't care that much about getting to the Faroes—or Iceland either." Kell is talking now, looking at Steve as he speaks. "For me it's the weeks and months of anticipation, and the satisfaction afterwards—of knowing you've done something challenging and difficult. I've sailed offshore—we all have. It gets long and tedious—days and days of wet sleeping bags and canned soup. But I've never sailed all the way across, and later, when it's all over, I want to be able to say I've done that. I guess, for me, the trip is going to mean one more notch in the gun handle—a big notch—one that I've wanted to put there for a long time."

Kell and I have been friends since our college days, and with him perhaps more than with any of the others, I feel I can read between the lines and translate his metaphor. In school Kell was a natural athlete and a fierce competitor. Winning was important to him because it always had something to do with who he was. When he sailed, he raced—and when he raced (as he did on a national level for several years), he did so with a ferocious concentration, and won.

Seven years ago Kell became the co-founder and part owner of his own computer software business, which is flourishing. In his spare time he has become an accomplished horseman; lately he has been training and riding in endurance events, the longest of which are grueling one-hundred mile rides. He has won several such events in the especially strenuous "cavalry" class, riding the events solo, without any aids to horse or rider.

A year ago Kell ran a marathon and finished. He didn't win. At forty-three years old there was no reasonable chance of winning. But he set a personal goal that would test the limits of his ability, and he achieved his goal. Some of the "notches" in the gun handle have thus become private and personal accomplishments—but they are notches nevertheless, and Kell has always found the wellspring of his tremendous energies in the pursuit of that next "notch."

Andrew looks at Kell with an expression of transparent bewilderment. "You make it sound so mechanical," he says. "Like passing a test for your next merit badge. But it's not that at all, at least not for me." He turns to Steve and seems to struggle to find the right words. "I just want to be out there, to feel the power of it all, the size, the incredible size of things. I want…." He opens his arms in a sweeping, wordless gesture and never finishes his sentence.

Andrew is seven years older than Steve. He has already sailed his way through the rites of passage and into young adulthood. He has worked on a charter schooner, raced to Bermuda, sailed a delivery north from the Bahamas, served as mate several summers on a private yacht. Next month during his January break from college, he will be skippering the delivery of a 32-footer from Beaufort, North Carolina, to the Virgin Islands. He is an addict who craves the sea experience—for its romance, its beauty, its challenge, its simplicity. "A yearning for something genuine in a world of trivialities"—these are partly his sentiments, partly mine. They seem to fit.

And what of my own reasons for wanting to do this voyage? I have been listening to each of the others as intently as Steve, although for a different set of reasons. I will add little tonight to what has been said. I've already sensed in each of the others' motives some part of my own. I want to sail with John for the adventure, with Kell for the feeling of personal achievement, with Andrew for the affirmation of something beautiful and simple—and with Steve, too, if he should decide to come, for the rite of passage into a new sense of self-reliance, competence, and mature adulthood.

A phrase from one of my favorite writers—Joseph Conrad—comes to mind, and this one I do mention to my shipmates. Conrad often wrote of "the fellowship of the craft," and by this he meant two things at once. The "fellowship of the craft" is a universal and unspoken bond among all those who have ever practiced the "craft" of sail and shared its secrets. But it is also the bond of intimacy that forms among the crew of any particular "craft" during the course of a single voyage. It is this second kind of fellowship, born of interdependence, trust, and mutual respect that I will also seek on the way across next summer. I'm certain that the challenges and rewards of a solo voyage are tremendous; but for me, the rewards of a comradery of shared risk and shared accomplishment will always seem far greater.

I think again about the Ithaka poem. Ithaka is important because the goal (however arbitrary) serves to set the whole process in motion. Without the goal, the voyage will never begin. And so I'll continue to dream of the Faroes—fog-shrouded, bleak, and treeless. Their headlands are dangerous. The currents and overfalls that race around them are often impassable. But the Faroes are the catalyst for a process that will last three thousand miles. They will give us the voyage. And that will be something; perhaps that will be enough.

At the end of every successful offshore voyage comes the bittersweet moment when the ocean experience is over and the voyager must reconnect with life ashore. The following piece recounts the feelings of a group of sailors who have crossed an ocean for the first time and have found themselves at the far end of the world.

2

Landfall Faroe Isles

A T EIGHT O'CLOCK SEVERAL RAGGED patches of blue invade the sky to the north. As they draw nearer, the wind drops to just under gale force and loses some of its bite. The color of the sea beneath long, breaking crests changes from aquamarine to dark green. The waves become steeper and more confused than they were last night, for we are now sailing up onto the shelf water of the Faroes Bank. The pelagic birds—shearwaters and fulmars—that have accompanied *Brendan's Isle* for two thousand miles since she left the Gulf Stream are mostly gone. In their place are blackback gulls and kittiwakes that soar under the lee of the reefed mainsail, just a few feet above the helmsman's head.

John and I share watch duties for the next four hours. He takes the first trick on the helm until nine o'clock, while I stay parked in the companionway under the lee of the spray dodger. Then he climbs below to thaw out while I take over the wheel for the next hour. At latitude sixty-one degrees the north wind is cold—but this morning I hardly notice. I am concentrating on the horizon ahead, searching for the first glimpse of land after twenty-three days at sea.

The stormy weather of the past week has taken its toll on all five members of this crew—yet there is a feeling of quiet anticipation this morning that can mean only one thing: imminent landfall. At breakfast Andrew reported a series of broken radar echoes that indicated less than thirty miles to the south cape of the island of Sudhuroy. Now, an hour and a half later, I squint at the horizon of cresting seas to the east and imagine for the tenth time that I see a thin line of hilltops jutting up among the waves. Mirages—only mirages.

What will these islands be like, I wonder. I know from the only large-scale chart of the region that the western coasts are bold, rising two thousand feet and more straight out of the sea. We all know there will be strong currents at the capes—we have been taking turns studying the set of current tables that I was able to purchase by mail from Denmark before we set sail from the States. These tables are organized by means of a series of hourly sketch charts that are keyed in to the lunar tables in the Nautical Almanac. But the text of the tables is in Faroese, a language evolved from ancient Norse during the time of the Viking settlement of these islands over a thousand years ago. Not even the Danes, who own and govern these islands, can easily understand the Faroese language.

We know there will be millions of nesting birds and tens of thousands of sheep on the hillsides. And we know that there will be no trees and very little agriculture. But what will the small communities of people be like? Will there be roads or cars? Will there be television or telephones? There is almost no available literature on the Faroes written in English. During our preparations for the voyage last winter, in fact, most of our American friends had no idea that such islands even existed. "You are sailing *where*?" they asked. "What do you want to go *there* for?"

I've been repeating the same question to myself all morning. Last night as I lay in my bunk, listening to the shriek of the wind in the rigging and the sound of water rushing past the hull, I found myself regretting that this passage had to end. This morning at breakfast as my shipmates and I shared a pot of oatmeal around the galley stove, I found myself apologizing to them. They had all worked hard without complaint. They'd sailed in the wind and cold for thousands of miles—and for what? A rockbound coast and current races. Barren hills without vegetation. Islands so insular and remote that their sparse population doesn't even speak a language that anyone else in the world can understand. A society of simple fishermen and sheep herders. A prohibition state, from all reports, without a single bar or public house to entertain a thirsty sailor.

Andrew climbs the companionway ladder near the end of my hour at the wheel and announces that his radar plot now indicates that we are closing in on fifteen miles. "Can you see anything yet?" he asks.

I squint again at the horizon ahead and shake my head. "Only the tops of seas."

The landscape of the Faroes, even from several miles off, is breathtaking.

He climbs up another step, craning his neck so that he can look around the edge of the spray dodger. "Oh my god," he mutters. "My god they are so BIG. They are so GODDAM BIG. THEY ARE GODDAM GIGANTIC!"

I let my eyes follow in the direction that Andrew is pointing, perhaps fifteen degrees off the horizon, and at last I see them: a line of jagged mountaintops, barely distinguishable in the haze. I had simply been looking in the wrong place. Somehow I had never expected them to be so large. Their surreal shapes slice across the sky, more like an illustrator's fantasy than solid land.

Andrew hollers down into the cabins, then back at me, then down into the cabins again. I sail for several minutes, finding it difficult to catch my breath, feeling as if someone has just clubbed me in the chest with a huge blunt instrument. As we continue our approach, the land resolves itself into darker and darker shapes. Soon our three shipmates also climb up on deck, and Andrew stops his incessant flow of garbled superlatives. For a time we sail in silence, all gazing at the strange silhouette growing in the sky ahead. Then, as if a dam had suddenly burst open, the cockpit of our little sailboat erupts into a chorus of cheers.

Brendan's Isle stands well offshore as she reaches northeastward, doubling the cape south of Sudhuroy half a dozen miles off, then motor-sailing another ten miles up the east coast of the island. The landscape, even from several miles off, is breathtaking. The island rises in a series of pinnacles to fifteen hundred feet and more. In the sunlight the colors of the land are black and emerald green, and the shapes are otherworldly.

There is nothing soft or round about the Faroes. They are geologically "new" land, part of the mid-Atlantic ridge and rift system formed by the radical folding and faulting of rock that has been pushed up out of the sea floor. Headlands cantilever out over the water; ridges slice across the sky like razors and rise into sickle-shaped peaks. In the valleys and coves that punctuate the coastline, several fishing villages appear miniaturized by the massive scale of the land around them.

In the late afternoon we enter a long, narrow gash in the rock, identified on the chart as the Vaagfiord. We proceed into this estuary for another five miles, all the way to its head, mooring at last alongside a stone fishing quay in the village of Vaagur. We have seen no other yachts anywhere on our approach, only a few double-ended rowing craft and a handful of small diesel-powered fishing boats.

Within minutes it becomes obvious that the arrival of a sailing yacht is something of an occasion in Vaagur. The quay fills with curious townspeople—several greet us in English. When they hear that we have sailed directly to their village all the way from America, we become instant celebrities. One man sends for the mayor; another offers to go find the customs agent. A third introduces himself as the principal of the local high school and invites us to use the school's swimming pool and showers. Cars arrive and park at the end of the quay. Children gather in little groups and stare. An old fisherman asks about the weather on the crossing. A young man wants to know why we chose Vaagur as our destination.

I try to explain that Vaagur actually "chose" us. We had originally intended to make our landfall sixty miles farther north, I tell him, nearer to the capital city of Tórshavn. But a day and a half ago a northerly gale forced us to change course and bear off to the south. Thus it was that Sudhuroy,

Some of the houses are topped with sod roofs.

the "south island," became the most logical destination. "You are the first Americans ever to come into Vaagfiord," he says, "the first yacht to arrive this summer from *anywhere*."

There seem to be plenty of roads and automobiles in Vaagur. I soon discover that there are televisions and telephones, too. But I can't help feeling that we are a bit like Cook or Magellan or Drake—explorers in a strange new world. I am quite unprepared for the novelty we seem to have created and for the enthusiastic welcome that continues unabated.

As we sit in the cockpit of *Brendan's Isle* talking with members of the welcoming committee, a sleek wooden pulling boat passes close astern. The boat is a lap-straked double ender with a sweeping sheer that rises dramatically at each end—obviously a close relative of the ancient Viking longboat. John, who is a boat builder by trade, comments on the lines and apparent ancestry of the craft. Andrew seems more interested in the crew: seven lovely Faroese girls, one at each of the six heavy sweeps, with a pretty young coxswain handling the rudder astern. They are all very

A Faroese women's rowing team pays a visit to Brendan's Isle.

blond and, as Andrew repeatedly observes, wonderfully healthy-looking. We are told that this is a local women's rowing team, practicing for an inter-island regatta soon to take place near the village of Tronsvaagur in the central part of the archipelago. As the rowers pull up the fiord into the cold wind, they are dressed in heavy white turtleneck sweaters with large geometric designs, a standard part of the summer wardrobe for many of the islanders. A mile to windward they turn the boat, pull off the sweaters, and sprint in shirtsleeves down past the town. All five sailors on the American yacht cheer with lusty enthusiasm as they pass.

At ten o'clock in the evening the sky is still flooded with sunlight. The hills behind Vaagur rise in steep ascent almost a thousand feet. A shadow cuts diagonally across them halfway to the summit, tracing the shape of a mountain on the western side of the fiord. Above this line the sunlight floods across vertical flutes of basaltic rock, casting alternate bands of shadow and light into high meadows carpeted in mosses and lichens and summer wildflowers. Several small streams dance down the rock on either side of the fiord and lose themselves in the green pastureland that lies just above the roofs of the town.

The town itself is arranged in streets carved in horizontal stripes into the hillside. The buildings are small and neat, painted in bright pastels: pink, blue, orange, violet, green, yellow, turquoise. Some are trimmed with bright colored borders around the windows and doors. A few are topped with sod roofs.

As the hour grows late, the shadows above the town creep imperceptibly toward the summit. The sky shimmers orange at the edges. Overhead, it is filled with pink mares' tails that swirl in complete circles, reminders of the strong winds that are still blowing out beyond the headlands. The date is only five days past the summer solstice, and in these latitudes there will be no actual darkness, only an hour or two of twilight in which one may still easily see to read or write in the cockpit.

No one aboard *Brendan's Isle* seems tired, even with the long day of sailing we have recently concluded. The quay is still filled with visitors. We invite the school principal and the mayor aboard, along with several others who have come bearing gifts of local foods, and we share a drink and a light supper. "Do people in the Faroes never go to sleep?" I ask. "Not at midsummer," one man exclaims. "We sleep in the winter."

Kell and Mikey decide after supper to climb the mountain behind the town. Andrew walks into the village, perhaps in hopes of encountering a few stray members of the local rowing team. One by one, our visitors bid us good evening and take their leave. John and I stretch out in the cockpit to watch the light rise again toward dawn.

My vague misgivings about making landfall in this place have long since disappeared. I am already feeling a certain remorse, in fact, for having allowed only five days to sail in these islands before having to push south for a planned rendezvous with Kay and our son Steve in the Shetland Islands. Is it just the inevitable excitement of a landfall, *any* landfall, that makes the people here seem so hospitable and their island so beautiful? I ask John the question out loud, then glance at the bloodied leg of raw, air-dried mutton that hangs on a short length of rope beneath the stern pulpit and the half-eaten slab of whale blubber that sits on a plate near the companionway, both gifts from our evening guests. No, it is obvious that we have sailed into a different world—one that I am eager to explore and learn more about during the few days that remain to us here.

*There is an ancient Hindu saying from the Rig Veda that
nothing is comprehensible except by virtue of its edges.
Maybe this is the reason some sailors journey so far from the
crowded centers of society to seek the boundary zones of life.
Maybe it's easier to comprehend our world—and ourselves—
from the perspective of these remote borders of experience.*

3 Searching for the Edges

WHEN I STARTED LOOKING FOR a group of young people to sail with
me aboard *Brendan's Isle* on her first summer's voyage to Labrador,
I knew that I wanted the enterprise to become something more than a
mere sight-seeing expedition. I hoped it might also become a journey of
the spirit. As such, it would require a great deal more than a strong boat
and a challenging cruising objective. It would also require a group of adventurers in search of the edges.

Not many sailors have the luxury of choosing their shipmates as carefully as I was able to do for this voyage. As skipper, I had corresponded during the previous winter with nearly forty applicants for the five available
berths. Afterwards, I personally interviewed more than a dozen of them.
The final crew—Beth Hawkins, Tim Clark, Scott Harris, Amanda Lake,
and Kirk Fitzsimmons—were all in their twenties, all strangers to one another, all at various stages of becoming adults in a complex world. Tim had
sailed with me the summer before from Bermuda to Newfoundland and
home again. He was, by virtue of that voyage, the veteran blue-water sailor
of the group. Amanda had once sailed on a delivery from Florida to New
England. The others were either small-boat sailors or novices, none with
offshore experience.

I should point out that I was well accustomed to sailing with crew members of all abilities, having directed a sail-training program for teenagers
for five years aboard the 60-foot schooner *Dawn Treader*. Since returning
from two years of European cruising aboard *Brendan's Isle*, I'd started sailing summers to Canada with older crews—not in a formal instructional

program but still a form of sail-training, I suppose. My crews have all been recruited by personal invitation, and they've all sailed as friends, sharing expenses. A number of them have been former sail-training students from the *Dawn Treader* days.

I knew for this upcoming voyage to Labrador that sailing experience would not be the most critical factor in choosing the crew. I was looking for something far more difficult to quantify than the miles they had sailed or the number of knots they could tie. I was looking for something ineffable—something interior.

Scott may have spoken for all the others one lazy afternoon along the coast of Nova Scotia as he and I shared a watch in *Brendan's* cockpit. He had graduated from college just a month before, and he was talking about one of the reasons he had decided to sail on this voyage. "I just need a little perspective," he said. "Too many of my classmates seem to be following some prescribed formula for success without really understanding where they are going or why. I need some time to be sure I know myself and to sort out what's important." There was no other sound when Scott finished speaking but the hiss of the wake and the creak of *Brendan's* rigging.

"You need to feel the edges," I said.

What are these edges, you might ask? And how do they affect those who travel on a long voyage to a distant coast?

Imagine running before a fresh southwest breeze in the Cabot Strait the first night out of Cape Breton Island. *Brendan's Isle* is underway on a three-day dash down the Gulf of Saint Lawrence to the Strait of Belle Isle and beyond—to the southeast coast of Labrador. The dark loom of Cap Anguille lies ten miles to the east. The green crew struggles to learn how to steer downwind as the big cutter rolls from gunwale to gunwale under a double reef and charges like a runaway train into the night.

On deck it is cold. The noise of wind and breaking seas that climb up from astern makes it impossible to talk. Down in the cabins there is the soft light of oil lamps and the glow of the radar screen, the smell of hot bread and shepherd's pie warming in the oven, the muffled sound of water rushing past the hull. Amanda is busy dressing for her trick at the helm: triple layers of winter clothing, wool watch cap and bright red mittens, boot liners and rubber boots, safety harness strapped around her foul-weather suit. She slides open the companionway hatch and reaches up to clip her tether to one of the jack lines.

A constant parade of icebergs migrates along the bays and inlets of the entire region.

A few seconds later Kirk appears at the hatch and climbs below. His face is flushed from the cold. His eyes betray the intensity of the hour he has just spent at the helm. After he strips off his heavy deck clothing, Scott hands him a cup of steaming hot chocolate and a bowl of shepherd's pie. Beth sits curled up at the dinette table in the main saloon strumming a guitar.

Kirk begins to talk—too loudly. I look up at him from my seat at the navigator's station and notice the emotion still written in his face. He glances at me, then at the others in the cabin, and he stops in mid-sentence. "Sorry," he says in a half-whisper. "This is such a...different world down here."

In this moment my shipmate may have experienced one of the most acute boundaries that any of us can ever know—a moment of heightened perception in which he is caught unawares with his pores wide open. What he must try to cope with is the border between being in control and letting go, between the noise and confusion of the deck and the quiet ambience of the cabins, between the solitude of an hour alone with nature and the interdependence of shipmates.

In a moment such as the one Kirk has experienced, a window opens to understanding. The responsibility that you bear to your shipmates and to the safety of the vessel is never as apparent as it is the instant you give over the helm to the next watch. Now you are the one who must trust the instincts and attentiveness of another sailor. As the night closes in around the boat and you withdraw into the subdued lamplight below, you experience a role reversal that dramatizes a fundamental truth about what it means to be a person working and living within a group.

Some of the edges, the most obvious ones, are not so much interpersonal as they are geographic. They have to do with the places you sail. Labrador is an edge of this kind. The harsh sub-Arctic climate of this region is magnified by the cold-water current that sweeps southward along the coast from far above the Arctic Circle. Even in the high summer months of July and August, the prevailing winds move the pack ice in from the sea all along the northern reaches. Meanwhile, a constant parade of icebergs and bergy bits migrates along the bays and inlets of the entire region.

The interior of this land, however, is the North American continent—three thousand miles of tundra, boreal forests, and grasslands, all the way to the Canadian Rockies and beyond. The coast is thus a boundary between warm continental air masses and the near-freezing waters of the Labrador current—perhaps the most dramatic meteorological interface in the northern hemisphere. The weather here is bizarre.

One morning *Brendan's Isle* breaks out her anchors in a sheltered cove several miles inland and sails before a light westerly to the mouth of a fiord called Ship Harbor. Here she meets a wall of fog. The wind dies and the air temperature drops thirty degrees. An hour later comes thunder and a black squall, with wind gusts of fifty knots and hailstones as large as golf balls. Then in another hour the sailboat is becalmed again on a glassy sea that undulates with the remains of a distant swell and reflects the dark blue of a cloudless Canadian sky like an imperfect mirror.

Icebergs punctuate the horizon in every direction—searingly white—backed by pink granite headlands to the west. To the east appears a shimmering semi-circle of black that looks like more land. But every person aboard knows that there is no land to the east—only two thousand miles of open Atlantic. We stare at this dark mirage, and slowly we come to see the spectral images of distant bergs, upside down, that lie many miles beyond the rim of the horizon. *Brendan's Isle* floats in an Alice-in-Wonderland world where the edges of reality itself have become transformed.

The effect that this strange geography has on *Brendan's* crew is more profound than you might expect. There are few familiar touchstones here to remind us of home. The sea is full of ghosts. The land is barren and nearly empty of population, with no buildings, no navigational marks, no

Reefed down and well under control as Brendan's Isle *enters the Bay of Islands, western Newfoundland*

distinguishing points of any kind to differentiate the redundant shapes of headlands. The kaleidoscopic patterns of our lives at home stand out in vivid relief against such a backdrop. Here it seems easier, somehow, to sort out the things that are important from the things that are not. Conversations in *Brendan's* cockpit become more reflective and philosophical. We talk about the health of the planet, global economics, feminine values, the evolution of the cosmos, geologic time.

The only permanent human population in this land is concentrated in a few dozen small towns and villages scattered along six hundred miles of deeply crenellated coast. *Brendan* visits one of them, a village called Mary's Harbor, with a population of some seven hundred souls. Later that evening, Beth writes the following observations in her journal: "The coast of Labrador is not meant for people. Nothing has the feeling of permanence here. The tiny prefab houses are perched tenuously on the rock, as if they could be packed up and moved away again tomorrow.

"Just beyond the town everything ends. There are no more buildings, no roads, no electric power lines, nothing that even hints at human imprint. I feel like here, at the edge of the town, I can understand what it must be like to live in this unforgiving land."

Beth makes friends easily with the people we meet. One evening, in a small fishing camp near the outer coast, she stands on a pier while a man fillets two large codfish (a gift for *Brendan's* crew). While she waits, she talks with the children. Later she writes: "It is difficult for us to understand each other's English. We are becoming friends by means of gestures and broken sentences. One child asks his mother how he can get red hair like mine. Another is amused at the size of our tiny rubber boat and six-horsepower outboard engine. He tells me proudly that his father's engine is *forty* horsepower...."

"Life is so simple here," Beth continues. "It makes me wonder why people seem so much more selfish back in the land of plenty."

During a journey such as this one, there are also emotional edges that test and sensitize every member of the crew. On windless days we are cabin-bound because of the swarms of mosquitoes and black flies that populate the tundra and follow us back from the shore. It is madness to try to do anything on deck during these times. Each crew member must then turn to himself to find the inner resources to cope with the silence of the cabins and the hours of waiting. Kirk weaves a hammock out of string. Scott bakes fresh bread. Tim, Amanda, and Beth play an endless tournament of triple solitaire. I write and read.

Several members of Brendan's crew capture a growler for the sailboat's ice box.

An hour later *Brendan* is underway in thick fog, and in place of tedium come anxiety and danger. The small slabs of floating ice, called growlers, are invisible on the radar, so the crew must alternate in the cold and damp as bow watch. The coastline is difficult to decipher. The charts are poor. And there is little chance of finding someone to help if we lose our way or strike some uncharted ledge.

One afternoon on the long windward beat homeward, *Brendan* fetches up under the lee of a tall headland on her way into the Bay of Islands on the west coast of Newfoundland. Here she encounters a series of blow-me-downs that pounce on her and knock her onto her beam ends, first to windward, then to leeward. I call for the crew to harness up. Scott and Tim must make their way to the foredeck to gather in the working jib. The others must retrieve a broken batten and tie a triple reef into the mainsail as I run off in a narrow channel. I can see the strain written in each of their faces. There is no room for error here and no time for hesitation. This is a test of the most serious magnitude.

In the bay itself the sky is black, swirling in katabatic downdrafts from the tops of two-thousand-foot peaks. The wind accelerates in a Venturi effect and whips the surface of the water into a lather of solid white. But now *Brendan* is properly canvassed and well under control again. She bears off and races before the tempest under the dark shadow of the land. I can feel the pride, the relief, the pleasure that every one of this crew feels as they gaze upward at the awesome scenery.

Another edge. This one having to do with competence and the knowledge that now we have come together as a crew and proved ourselves good seamen. What a difference between these sailors and the ones who were so tentative and uncertain of themselves only six weeks before.

The Bay of Islands is still thirteen-hundred miles from that other bay, the Chesapeake, where this journey began. There are many nights of gales and tedious days of calm to test *Brendan* and her crew before we can rest in the accomplishment of another circle. At the conclusion of this process, another life will be waiting for each of us. But it will not be quite the same as it was when we left two and one half months before, for in between have come the edges, changing the way we understand ourselves and the way we think about the world.

Why do we act so foolishly? Because we are blind and deaf. We cannot see or hear what is right before us. If we did, we would not carry out this assault upon the Earth.

—Thomas Berry, *The Dream of the Earth*

4   The Wreck of the *Braer*

O N THE MORNING OF JANUARY 6, 1993, news media around the world carried the story of a shipwreck and major oil spill in the North Sea. An eighteen-year-old oil tanker, the *Braer,* of 89,700 tons burden, had driven aground the previous afternoon. The ship was in imminent danger of breaking up on a remote and sparsely settled coast near the Bay of Quendale, three miles from the southern tip of Mainland, Shetland Islands.

After reading the report of the spill in a local newspaper, I found myself staring at the accompanying photograph of the ship, its superstructure raked by storm seas, a large oil slick spreading from its center toward the nearby coast. There was something disturbingly familiar about the setting of this photograph. I looked again and realized that I knew the place where the drama was unfolding. I had once made landfall on this coast—had looked down into this very bay, in fact, from the cockpit of *Brendan's Isle.*

My original encounter with this place had occurred one foggy morning in July, 1984, on the final leg of a long trans-Atlantic crossing. My shipmates and I had been watching ahead that morning, waiting for the loom of Fitful Head to resolve itself into something solid on the horizon. As we'd slowly closed with the shore, we'd watched a pair of gannets framed against green-tinted seas and we'd laughed at the clumsiness of puffins swimming with their young on the surface of the swell. Later, with the gray shapes of Lady's Holm, Horse Island, and Scat Ness rising out of the mist, we'd listened to the complaints of seals basking on the rocks and we'd counted seabirds wheeling in the air above Sumburgh Head, like blizzards of summer snow.

My shipmates and I had been privileged that morning to bear witness to a natural place at the edge of the modern world—a place that most people never see. We'd spent a few careless hours with the inhabitants of that place, listening to their voices and feeling their solitude—once upon a time, when their world was still whole.

The newspaper's account of the wreck of the *Braer* opened with a description of "oil gushing into waters teeming with marine and bird life." One sentence early in the report mentioned an imminent threat to "thousands of puffins, loons, gulls, and eider ducks." But otherwise, the story suggested, things didn't appear that bad. It was nearly certain that all twenty-five million gallons of light crude carried on the *Braer* would spill into the sea as the ship broke apart. But, according to the Associated Press report, the fact that the crude oil was not thick would help minimize the damage. "Around forty percent can be expected to evaporate into the atmosphere and a further twenty to thirty percent to disperse in the very heavy seas," the report indicated, while "on rock shores the oil will be washed off and dispersed by the waves."

Evaporated into the atmosphere? Washed off? Dispersed by the waves? I was surprised at first, then annoyed, and finally angered by the article's almost casual tone.

At twenty-five million gallons, the *Braer* was carrying more than twice the volume of crude oil that the ill-fated *Exxon Valdez* carried when that ship struck a reef in Alaska's Prince William Sound in 1989. The news report, while stating this fact, compared the two events as follows: "That disaster [the *Valdez* spill] dumped thick, heavy oil into a relatively calm, enclosed body of water while the *Braer* was grounded in open seas... on the rugged shore of the Bay of Quendale, one hundred fifty miles northeast of the coast of Scotland."

This explanation would seem to suggest that the site of the *Braer's* grounding was in some "safe" location, many miles from the nearest threat to land or life. But I knew better. Three miles south of the Bay of Quendale is an area known locally as Sumburgh Roost, a dangerous tidal race where the current flows around the southern tip of Shetland at velocities that can carry oil-laden sea water long distances. Twenty miles north of Sumburgh Roost, on the east coast of the island of Mainland, is Lerwick, a town of ten thousand inhabitants and the largest settlement in Shetland. The same distance away on the west coast is the island of Hildasay, along with

Puffins stand in the mornings like sentries next to their nests.

hundreds of other islands and voes (as the long, narrow bays of this area are called). Twenty-five miles to the west is the rookery island of Foula. Twenty miles to the south is Fair Isle, a stark and beautiful area of small sheep crofts that is also one of the most important bird sanctuaries for migratory fowl in all of northwestern Europe. Based on the known distances that other major oil spills have traveled, all these areas would soon become candidates for serious damage.

The four-hundred acre island of Hildasay, some nineteen miles north of the shipwreck site, was owned for a time by my two good friends, John and Sandy Griffiths, so I know this place more intimately than most of the others. Hildasay is a deserted sheep croft with a small natural harbor, two miles across North Channel from the Mainland village of Scalloway. The area around Hildasay, like most of the coast of Shetland, is little changed from the way it might have looked a thousand years ago when Danish marauders first sailed their Viking ships to these islands and decided to settle here. Trees do not grow on Hildasay—only windblown tamarack, thistle, sea oats, and wild grass. The sheep—no longer present—have grazed the island to an Arctic nakedness, leaving its silhouette of low cliffs and mottled outcrops as a stark signature of the place for all to read.

Under a hill on the east side of the island are the stone foundations of the ancient croft. In the harbor below, protected by a shallow rock ledge and an islet named Linga, is a tidal channel where Mainlanders have built a commercial salmon farm. But except for the floating pens and storage sheds, there is no other sign of human habitation on the island. Instead, nesting along the western cliffs and in the grassy hills and on the shores of a tidal voe that enters the island from the southwest, are the birds. Puffins stand in the mornings like sentries next to their nests. Herring gulls and terns, guillemots and eider ducks, shags and great northern divers fish in the surf. Turnstones, oystercatchers, and curlews wade in the shallows at the end of the voe.

In the cold water that surrounds the island are codfish, sandeels, and sprat—fish stocks that have persisted in commercial numbers all around the Shetlands, sustaining a local inshore fishery for hundreds of years. Lobsters and velvet crabs populate the shallow rock ledges. Common and gray seals fish the Deeps and congregate on the boulders along the western shore. Sea otters bask in the surf and fish among the ledges of Middle Channel.

This, in any case, is how the island appeared when my friends used to camp here in a canvas tent during the summers of the late 1980s. But now, as soon as the prevailing southerly winds blow for a day or two, the crude oil from the *Braer* will creep north to fill the bays and channels, clogging the voes, coating the rocky shores and beaches with its greasy, toxic film, and causing most of the inhabitants of Hildasay to slowly suffocate and die.

The simple fact is that no place on Earth is "safe" for this kind of accident. No shoreline is so remote that it can somehow accept the impact of twenty-five million gallons of spilled crude oil without receiving a mortal wound. No body of water is so large that it can accept the trauma of a spill of this magnitude without changing its basic photosynthetic chemistry and its capacity, at least for a time, for sustaining life both above and below its surface.

There are many possible causes for an accident such as the grounding and eventual breakup of the *Braer*. In this case, one of the main causes may have been the ship itself. Built eighteen years prior to the accident, the *Braer* belonged to a generation of "budget model" oil carriers that were constructed of a single-layered hull with no interior tank walls to carry the actual cargo. Such carriers were typically built for about ten years' useful life before the riveted steel plates and other structural components of their hulls began to weaken under the stress of deep ocean service. At this stage the ships should probably have been scrapped or extensively rebuilt. Too often, however, the original operators chose to sell them to smaller operators who were willing to take the risk of running the weakened ships in exchange for large profits-per-voyage. "Coffin ships," these aged carriers have come to be called. "Floating time bombs."

Old age and structural infirmity weren't the only problems that beset the *Braer* on the day she was lost, however. When her diesel tanks ruptured in the rough seas, salt water invaded the ship's fuel supply, eventually forcing the shut-down of her engines in the narrow channel between Shetland and Fair Isle. Recognizing that his ship was now drifting out of control, the captain radioed the Coast Guard station in Aberdeen, Scotland, requesting that he and his thirty-four man crew be removed from the stricken vessel. The rescue was accomplished approximately five hours before the *Braer* drove onto the rocks under Garth Ness.

Shortly after the helicopter evacuation and "two hours before [the *Braer*] went aground," according to BBC reports, a salvage tug named *Star Sirius* arrived on the scene. A spokesperson for the tugboat's owner told BBC that there was no one on the tanker when the tug arrived. "If there had been somebody on the ship prepared to give the tug captain a wire or line, ...[he] could have connected up without any doubt and... this thing wouldn't have happened."

In response to criticism that the evacuation of the *Braer's* crew may have taken place too soon, the regional controller of the Aberdeen Coast Guard defended his organization's decision to act as they did. "Our first priority," he explained, "is the safety of life."

Our first priority is the safety of life. If only this were so. If only we, as individuals and as a society, were somehow willing to measure the meaning of our acts according to "the safety of life." If only the *Braer's* captain, or the Coast Guard controller, or the helicopter pilot, or the tanker's chief engineer, or the owners of the stricken ship were willing to make their first priority "the safety of life." If only the Associated Press were willing to speak out in advocacy for "the safety of life." If only we, the public, were willing to look critically at how we live, inspect our habitual patterns of consumption and waste, and act together for "the safety of life."

The problem, in the end, is not just how we build ships or how we move our oil from place to place or how and when we decide to evacuate thirty-four members of a stranded crew. The problem is how we think about ourselves and the Earth and how we may somehow contrive to share this planet with all the rest of the biological world that is also trying to live here.

I feel lucky. I've spent much of my adult life as a kind of sojourner in the silent world of sail, traveling to places like Foula and Fair Isle, Hildasay and Quendale Bay. I've been allowed to listen to the wind, to witness something of the fragile miracle of nature, to feel myself in touch with other living creatures, and to contact the rhythms of the planet. These experiences teach a kind of courtesy toward the Earth, if you are willing to pay attention. But they also carry with them an imperative that is hard to follow in today's world. They ask for witness. They ask for advocacy in the cause of "the safety of life."

Thomas Berry may have gotten it exactly right when he suggested that the root of our collective myopia about the Earth is our anthropocentrism, our habit always of thinking about ourselves as the predominate form of life, and our insistence on measuring the value of the rest of nature in human terms. "This anthropomorphism," Berry writes, "is largely consequent on our failure to think of ourselves as a species." What he means is that we've failed to understand (or have forgotten) how we fit into the scheme of things, how we are one (among many) species, how we are dependent on the health and prosperity of the whole.

In the weeks following the breakup of the *Braer,* the inshore fishery was closed until further notice in southern and western Shetland and salmon farms subject to toxins were condemned. Hildasay, according to eyewitness reports, lay surrounded for many weeks by a greasy black slick of oil.

An accurate accounting of wildlife killed as a result of the spill may never be possible, but estimates made a few days after the accident already indicated a death toll of thousands of shags, black guillemots, puffins, and eider ducks, hundreds of seals and sea otters; thousands of lobsters and velvet crabs, a pod of killer whales observed in the area, and many hundreds of turnstones, curlews, little auks, terns, oystercatchers, herring gulls, and great northern divers, among many other species.

"The safety of life" is under siege for the foreseeable future in the Shetland Islands. The safety of all life—including our own—is under siege on the planet until we can somehow find a better way to manage our affairs and to present ourselves, as Thomas Berry states, "in an evocatory rather than a dominating relationship" with nature.

Many who have voyaged aboard Brendan's Isle *have come to her first by answering an ad in the sailing press. Never before, however, had so many replied to an advertisement for crew as in the months before her voyage to western Greenland...*

5

Seeking Experienced Crew

DURING THE WINTER OF 1994, a short and somewhat unusual notice appeared in the classified pages of *Cruising World* magazine. The ad ran for three months under the general heading "Positions" and read, in part, as follows: "Seeking experienced crew for a summer voyage to Greenland. Must be over 21, compatible, fit, eager for the challenges of a journey to the ice. Contact Captain Mike Arms, P.O. Box 30, Cecilton, Maryland."

As of this writing, *Brendan's Isle*'s six-thousand mile Arctic adventure is over. The story of this voyage and of the sailors who made it has been chronicled in a book, *Riddle of the Ice*. But meanwhile there is another story to tell—equally compelling—about an even larger group of sailors who answered the ad in hopes of taking part in the adventure and who were, of necessity, left behind.

"Dear Captain Mike," wrote Ken, a freelance filmmaker from New York City. "As I listen to the honks and sirens outside my window, I pick up a copy of *Cruising World* and take a trip without going anywhere. Then, suddenly, I run across your ad...."

Ken was one of nearly one hundred people who answered the ad and volunteered to join the crew of *Brendan's Isle*. Some were content merely to dash off a few lines on the back of a post card. Others sent lengthy résumés of sailing experience, skill levels, training, and sea time. But most who answered chose to write more personal letters, describing themselves, their love of sailing, their hopes of voyaging to faraway coasts.

"Dear Captain Mike, I am a lawyer, a sometimes writer, a newspaper reporter, and an avid reader of Greenland lore. Ever since I read Rockwell

A portrait of Brendan *in the ice, western Greenland*

Kent's *N by E* as a boy, I have dreamed of the experience of making landfall on that stark, mountainous shore...."

"Dear Captain Mike, I grew up in Maine. My mother's side of the family were all deep sea sailors and lobstermen. My father was a ship builder. I've inherited from them a love of the sea and travel. After twenty years of work as a carpenter, I'm selling off my possessions, getting down to a truck, my tools, and a small sailboat. I am very interested in your voyage...."

"Dear Captain Mike, When I left my job in Montreal to work as crew on a 92-foot Australian square-rigger, I wasn't sure I'd be accepted as a woman in what has traditionally been a man's world. But now, a month after returning home, I know I am as capable as any man. It is snowing today in Montreal, and I find myself thinking about other, more challenging journeys—maybe even a journey to Greenland and the land of the midnight sun...."

It might seem logical to assume that the people who answer such an ad will all fit rather conveniently into a few predictable categories, but such an assumption is far from the truth. For every starry-eyed co-ed who sat down to pen a response, there was a professional merchant seaman or a hard-nosed tugboat captain. For every wet-behind-the-ears novice cruiser, there was a seasoned blue-water sailor with fifty thousand miles under his keel. For every unattached thirty-year old with a wander-lust, there was a fifty-five year-old career professional with spouse and family.

Among those who answered the ad were a college championship football player, two recently graduated triathletes (one male, one female), several middle-aged marathoners, and a sixty-two-year-old Greek-American alpine mountain guide. There was a salmon fisherman from Alaska who had been shipwrecked recently near Kodiak Island, a U.S. military Arctic survival instructor who had once traveled by foot across the Greenland ice cap, an officer in the Spanish Navy who had served ten years as a professional race-boat navigator, a Canadian boat builder who had sailed to seventeen foreign coasts and now yearned to see the remote areas of his own country, and—perhaps most impressive—a thirty-four-year-old Virgin Islands charter skipper, twin-engine pilot, scuba instructor, skier, backpacker, kayaker, painter, writer, photographer, and former gold-medalist skater, who mentioned at the end of her letter that she was also—incidentally—a woman.

Last winter was not the first time I'd ever placed an ad in a sailing mag-azine looking for crew. But it was the first time I'd ever received such an overwhelming number of responses—and the first time, too, that so many answers had been characterized by such undisguised passion for adventure.

At first I couldn't understand why. Greenland is not a destination that interests most sailors. The water is cold and full of ice. The coasts are mostly uninhabited and poorly charted. The voyage north from Newfoundland begins with a transit of some of the stormiest seas in the summer Atlantic. The voyage home finishes with a windward passage of nearly two thousand miles.

A few of the letters I received might have been naïve and starry-eyed, written by people who simply lacked the experience to understand the kinds of risks and challenges they would be letting themselves in for. But most were anything but naïve. Young and old, male and female, North American, European, African, Asian, Australian—the vast majority were sailors with long and varied experience, who understood precisely the kinds of risks and challenges they would be letting themselves in for. It was almost as if the idea of voyaging to Greenland had struck a universal

Liz and Amanda... two of the best who have ever sailed aboard Brendan's Isle

chord—capturing the spirit of challenge and camaraderie and contact that so many sailors seek—until it came to characterize, for these correspondents at least, the archetypical cruising objective.

The only problem I encountered in reading the letters of these hundred-odd sailors was that I didn't have enough room aboard the boat to take them all. After the mate and myself, there were places for only three more crew members aboard *Brendan's Isle.* Choices had to be made—sometimes arbitrarily—based on factors such as geography, age, sex, work experience, educational background, and my own hypothetical notion of what an appropriate mix might be in a group that would have to sail and live together harmoniously for the better part of four months.

In the end, this meant that scores of excellent shipmates did not travel to Greenland aboard *Brendan's Isle.* Of all the people who did not join us, however, there was one, a forty-year-old wanderer of the world named Remus, whom I may have missed most of all. Originally from Lithuania, Remus was a man with a slender pocketbook and insatiable wanderlust who spoke and wrote in four languages. By some indomitable combination

Working together under pressure is an essential skill for every good offshore crew.

of planning, luck, and sheer determination, he had visited every state in the Union, every country in Europe, and every continent but Antarctica.

"Dear Captain Mike," Remus began, "This is big adventure to the ice. I'm dreaming about it long time ago. But problem—no money. I have only video camera when I'm traveling around the world.

"I'm very very brave man and also very extremely friendly. I'm very serious—no drink, no smoke—and very extremely intelligent. I'm sailed on the sea many long journeys, but on the sailboat only one time. It was 1991 June from Hawaiian Islands to the Canada on the 42-foot sailboat sloop. It was one of the most bad climate years 1991—two tropical storms hit us very strong, about twelve-force winds. I'm never forget this trip to the Canada!

"I have green card and passport, Captain Mike—no problem. If you very serious, I'm immediately come to Maryland to your house—so write soon. I'm wait to hear from you."

Sadly, by the time I read this letter, both the mate and the full complement of crew had already been selected and I had to tell Remus that he would not be coming to my house in Maryland—at least not to sail on this voyage. At the same time, however, I told him how impressed I was with his eagerness, his self-confidence, and his bravery in the face of physical hardship and danger.

The fact is that I'm deeply impressed by all sailors who say they would like to voyage along difficult coasts, across cold, icy seas, to places they know almost nothing about, with shipmates they've barely met. I'm astonished at the number of people who consider quite seriously doing something like this. It makes me realize that we still live in a world of people who value adventure, who are willing to take personal risks, and who are motivated to seek and to celebrate the beauty of this planet.

There were nearly one hundred sailors who, like Remus, wanted to voyage to Greenland but could not. Every one of them was eager, self-confident, and "very, very brave"—with varying degrees of experience at sea, to be sure, but with a universal desire to do something new and challenging. I like these people. Reading their letters has been a wonderful experience—one that has rekindled a sometimes flagging faith in our maligned and embattled species.

Part II

A Gaunt Waste in Thule

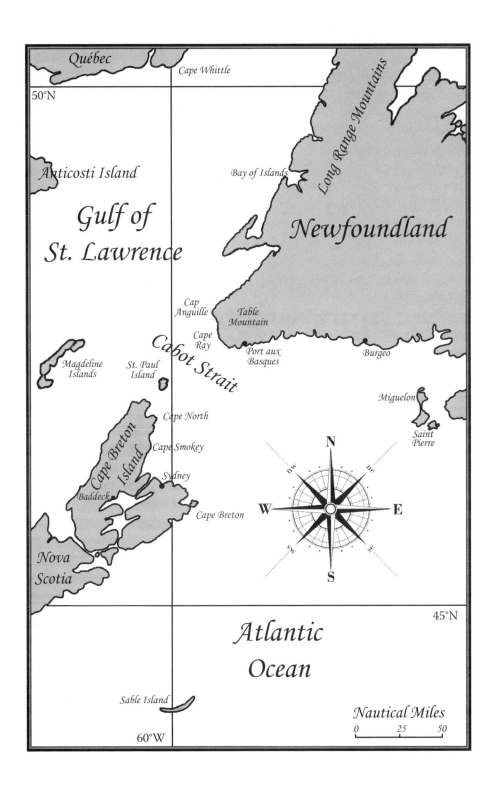

These are the waters of beauty and mystery... they fill the lodes in my cells with a light like petaled water, and they churn in my lungs mighty and frigid, like a big ship's screws. And these are also the waters of separation; they purify, acrid and laving, and they cut me off.

—Annie Dillard, *Pilgrim at Tinker Creek*

6   Waters of Separation

Between the highlands of Cape Breton, Nova Scotia, and the western capes of Newfoundland lies a body of water seventy miles wide, the Cabot Strait, that severs the world of North America in two. On the Nova Scotia side of this strait is a tamed and obedient land of paved highways and shopping malls, airports and trailer parks, souvenir shops and buses full of summer tourists. On the Newfoundland side is a stark and tortured coast, sparsely settled, with outports criss-crossed by footpaths, crooked, hand-built fishing stages, snow lying in patches in the July sun—a world so little changed from the way it was when the first Europeans settled here more than three centuries ago that it seems to have been cordoned off from the rest of North America, turned in upon itself, and frozen in time.

I am drawn to the Cabot Strait like an addict. I am mesmerized by the gray-green seas, the seabirds soaring in the sailboat's wake, the basking family groups of fin whales and humpbacks that appear without warning on the horizon. I am awestruck by the landmarks that one by one measure the progress of our little sailboat's crossing: the hulking presence of Cape Smokey, the loom of Saint Paul's Island, the layered ribbons of fog on Table Mountain, the black silhouette of Cap Anguille.

I have sailed from west to east across the Cabot Strait many times— on the way to western Newfoundland, or the Saint Lawrence Gulf, or the Strait of Belle Isle, or eastern Labrador, or the ice fields of western Green- land—and every time the experience has been the same. No matter how far my crews and I might have traveled to get to this point, the Cabot Strait

We point our bows out past Aconi Point and into the gray-green waters of the Strait.

has always presented itself as the place where the real voyage begins. Every mile that has come before feels like preamble: familiar, safe, predictable. The Cabot Strait marks the end of preamble—the end of the familiar.

On each of these voyages, the last port of call for our sailboat before entering the Cabot Strait has always been Baddeck, Nova Scotia, a small village on Cape Breton Island, nestled among gentle green hills on the northern shores of the Bras D'Or Lakes. Here, in the relative safety of a landlocked area of inland channels and thoroughfares, the sailboat and her crew pause for a few days to top off water and fuel, replenish fresh food supplies, and replace or repair whatever items of gear may have been lost or broken during the eastbound passage from the Chesapeake Bay. As the days of preparation draw to a close, a feeling of impatient expectancy grows among the crew. Finally, on a morning when the wind turns fair and the tide begins to run out toward the sea, we cast off the sailboat's mooring lines and point her bows east once more. Moving under power, we proceed down a twenty-mile fiord called the Great Bras D'Or, under a highway bridge, through a narrow tidal race at Aconi Point, and out into the gray-green waters of the Strait.

Even after half a dozen such passages, I still feel a twinge of fear each time I encounter this place. The reasons are obvious for anyone who sails here. The water near the coasts swirls in odd currents and overfalls. Fog banks as dense as night materialize out of nowhere. The west wind tumbles down off the highlands of Cape Breton, attacking the surface of the sea in katabatic blasts. The east wind brings a damp, wintry chill that feels like January in July. Both the south and the north winds rise to gale force and beyond as they are squeezed between the steep shores of Saint Paul's Island and the nearby headlands of Nova Scotia and Newfoundland, accelerating in a Venturi effect along the contours of the land.

The Cabot Strait is a wild and lonely place that cares not at all for the affairs of humankind. In this sense, at least, it is an ideal setting in which to begin a voyage, for even as it serves as a threshold, opening to whatever strange new worlds lie beyond, it also functions as a barrier, cutting you off from the familiar worlds that have come before. The experience of sailing here, drifting aimlessly in a gray cocoon of fog, or beating into a cold northeasterly, or racing wing-on-wing down the faces of following seas, inevitably serves as an invitation for the sojourner to ponder the big questions. Why have I journeyed to this place? Why do I return here, over and over again? What have I been seeking? What lies out there beyond the Strait that holds such a powerful attraction? And what is it that I am forced to leave behind and that feels so terrifying and so utterly liberating, both in the same instant?

For me, such questions inevitably lead back to a cluster of memories about the midwestern American suburb where I spent the first eighteen years of my life and about the boy-child who grew up there. The place was Shaker Heights, Ohio, a well-to-do community of large homes, manicured lawns, winding, tree-lined streets, white clapboard churches, fieldstone country clubs, red clay tennis courts, designer golf courses. It was a sort of utopian experiment, I suppose, an attempt to express in wood and concrete and stone that peculiarly American belief in the control of nature and the perfectibility of the human condition.

People were comfortable in Shaker Heights. We spent our lives wearing the symbols of our social predominance and our economic success: expensive clothes, fast automobiles, large houses, private backyard swimming pools. We surrounded ourselves with orderliness. The streets were cleaned by town employees once a week. The lawn cuttings and leaves and

John and I secure the rig in preparation for a boisterous crossing of the Cabot Strait.

weeds from the gardens were carted away by the professional yard crews. People didn't put their trash cans out on the street on trash day in our neighborhood. Instead, the trash men walked to the back of each house (to the garage that was, by special zoning ordinance, forbidden to be visible from the street), and carted the trash cans down to their trucks, then carried the empties back to the garages to place them back in hiding. (There was, one was encouraged to believe, no trash in Shaker Heights.)

The public parks in my town were accoutered with green wooden benches, concrete picnic tables, groomed cinder footpaths, little stone bridges spanning the creeks and drainage culverts. The woodlot behind my grammar school had no understory, only large, mature oaks and hemlocks, tulip poplars and red maples, elms, hickories, and horse chestnuts—a whole forest of trees that some enterprising local naturalist had labeled with bronze plaques affixed to each tree and engraved with each one's Latin name.

As a child growing up in Shaker Heights, I loved to do two things on weekend mornings when my time was my own. The first was reading books, an undertaking always considered a noble endeavor in my family and always permitted—even encouraged—by my parents. The second was wandering among the creeks and gullies of a large and mostly trackless area of undeveloped bottomland on the fringes of our town called the Shaker Woods—an enterprise considered dangerous by my parents and therefore just as emphatically forbidden.

On days when the weather was rainy or when the snow lay thick upon the ground, I would find a book, hide myself away in some quiet part of our house, and read until my eyes were sore. At first, the books I read were chosen by my mother. She was the reader in our family; her great Uncle "Billy" Phelps had been a noted teacher of literature at Yale, a man revered by my mother and transmogrified over the years into a family icon of superhuman proportion. As I grew older (and the ghost of Uncle Billy grew less omnipresent), I was invited to choose books on my own, and soon the titles I was reading began to fall into a predictable pattern: *Mutiny on the Bounty, Typee, Two Years Before the Mast, Kon Tiki, Byron of the Wager, Lord Jim, Count Luckner the Sea Devil, Fatu Hiva, Red Orm, Typhoon.*

On other days when the weather was pleasant, I would set aside the books and would fabricate a destination—a classmate's house, perhaps, or a trip with a friend to the movies, or a sports event at school—then I would set off on my own for a few hours of aimless wandering in the gullies of the Shaker Woods. Although I was not able to find the words at the time, I think

What lies out there beyond the Strait that holds such a powerful attraction?

I understand now the lure of those gullies. They appealed to me not just be-cause they were forbidden—although that was certainly part of the reason—but because they were untamed and trackless and wild-feeling. They were the only places I knew that had not been trimmed and pruned, edged and manicured, the only places without groomed cinder paths and little stone bridges and bronze labels with Latin names on the trunks of the trees.

The entrances to the Shaker Woods were never marked. Some lay hid-den at the far ends of the public parks; some opened inconspicuously at the edges of highway bridges; some lurked in the backyards of houses that sat in rows upon the heights. For the first few steps as you left the world you knew and moved into the undergrowth, you might have been able to follow a lightly worn path. But as the foliage grew denser, the path inevitably dis-appeared. And then you simply followed your nose, down through thicker and thicker areas of leaves and briars and towering trees, into a deep ravine with a meandering creek at the bottom.

There was beauty down in the gullies, tangled and dark. There was an endless fascination of living things: tadpoles and minnows, frogs and tur-tles. There were plants you would never see in the manicured gardens on the heights: May apples, crocuses, wild violets, jacks-in-the-pulpit. There were treasures scattered on the ground: tiny flecks of mica schist, gray slabs of slate, flint chips, dark red garnets and white marble pebbles.

For the little boy from the groomed and orderly world above, wandering in the world of the gullies was like wandering in a strange and exotic marketplace. I loved the jumble and complexity of that world. I loved its randomness. I loved its surprises: the dappled purple shadows of the deep forest, the sudden shafts of yellow sunlight at the edge of a clearing, the hissing of the stream as it flowed around an obstruction, the chatter of squirrels interrupted in their task of nest-building.

On some mornings the world of the gullies became overwhelming, so laden with new sounds and smells and colors and unfamiliar objects and alien creatures that I would become disoriented and lost. Which way was home? Where had I entered this maze? What direction had I been walking along the creek bed? Had I made my way through a culvert under an unknown highway? Had I followed an offshoot of the main gully toward another township far from home? I would stop in my tracks, and for a moment I would feel afraid. This was a dangerous place. Perhaps my parents had been right to forbid me to come here....

Then I would recognize a landmark, orient myself again, and set off to retrace my steps toward home. When I arrived at the edge of the woods, I felt like a conquering hero: Livingston emerging from the African jungle, Nansen returning from the frozen north. Flushed with pride and feelings of mastery, I would stride toward home wearing a torn trouser knee or a scraped elbow or a wet and muddied boot, and I would brace myself with stoical resignation for the inevitable punishment in store for me.

For reasons I now understand, the gullies of that Ohio childhood were always considered by my parents to be forbidden territory. In the years since that time, I've often wondered why the books were not considered forbidden territory as well—if not by my mother and father, then at least by the society at large. For like the gullies, the books, too, were subversive things, dedicated to a different god. They were windows into a world that did not exist in Shaker Heights, a world of voyagers and adventurers and roamers, a world of trackless ocean wilderness and untamed natural places that called into question the basic assumptions of much that was considered sacred in that comfortable and orderly world upon the heights.

Among the books I read during those years, two that I remember most vividly were tales told by young, would-be adventurers who dreamed of voyaging in their own small boats to unknown coasts. Both were presented as personal memoirs about actual voyages, and both began with dramatic accounts of sailing passages across the Cabot Strait.

The first, *Northern Light* by Desmond Holdridge, was a classic coming-of-age tale about three young men who decide, after their formal education is finished, to purchase a tiny schooner-rigged Tancook "potato-lugger" in southern Nova Scotia and sail her to the northern tip of the Labrador Peninsula and home again. Penned in the 1930s by the American leader of this trio, the book becomes a kind of odyssey of self discovery, a journey from innocence to experience, evoking both the stark beauty of the northern coasts and the sheer exhilaration of finding one's way along them without useful maps or other modern navigational aids. The book almost ends in tragedy in the Gulf of Saint Lawrence during a late October gale, as the three are forced to abandon their sinking vessel and accept rescue at the hands of a passing fisherman.

The second book, *N by E*, by Rockwell Kent, chronicles another small boat voyage, this one beginning along the coast of southern Connecticut and finishing (again in shipwreck) in a desolate and storm-wracked fiord in western Greenland. Written in a decorous and rhythmic prose that approaches poetry at times, the book describes a journey that Kent had undertaken when he was a young man (and before he achieved his later notoriety in the art world for his drawings and woodcuts). Several pages into the tale, he and his shipmates experience their first real night at sea as they set out across the Cabot Strait. In a voice filled with nearly equal portions of fear and awe, the young artist-turned-sailor describes his initial solo trick at the helm:

> *And as it darkens and the stars come out, and the black sea appears unbroken everywhere save for the restless turbulence of its own plain—then I am suddenly alone. And almost terror grips me, for now I feel the solitude; under the keel and overhead the depths—and me, enveloped in immensity.*

Even as a boy, reading in the quiet sanctuary of my parents' home in Shaker Heights, I was mesmerized by such accounts of the Cabot Strait and the bleak, wild coasts beyond. I had a suspicion that these places were not so terribly far away, and I promised myself that some day I would go there. I wanted to feel, with Rockwell Kent, the solitude of a stormy night at sea and the challenge and immensity of a grand and forbidding strait. I wanted to awaken, as Holdridge had done, in the company of "giant icebergs under our lee," and to witness, with Kent, "a grim land, shrouded in scud, steel gray against the low dark ceiling of the sky"—the island of Newfoundland as it rose on the sailboat's bow. I wanted to encounter a society of simple fishermen whose experience of the world was fundamentally different from mine, whose subsistence lifestyle and extreme isolation in this "uncultivated land" had formed them into "a people kind through necessity, wise in unworldly ways, virile, brave and good by the grace of God."

Years later, when the opportunity finally presented itself for me to sail to this place, I felt just like Holdridge and Kent and all the other naïve and wild-eyed adventurers that I had read about in books, for I, too, journeyed here as an innocent, unsure of exactly where I was going, uncertain of what I sought. I was confident there would be danger in such an undertaking, and risk—and indeed there have been, although they've been of a different order that I first imagined. I had expected challenges of strength and skill and physical endurance, and I was certain that with preparation and care my shipmates and I could handle these. But what I did not expect was the feeling, each time I returned from one of these journeys, of having been cut off, in some dark and unsettling way, from the life I had left behind only a few short months before. I did not expect, after a summer of exploring the Labrador coast or gunkholing in the bays of eastern Newfoundland or wandering among the ice flows of western Greenland, to come back home a stranger, staring wide-eyed at the most ordinary sights, recoiling awkwardly at every commonplace encounter.

In part, this feeling of being cut-off is similar to the feeling of anyone who travels for many months to a distant and unfamiliar land. When you return to the place you call home, you do so in a state of culture shock. Your defenses are down, and at least for a brief time, you're actually able to *look* at things as if you'd never seen them before. The routine, the normative, the ordinary—all take on a quality of strangeness. You feel like an immigrant, stepping out of another time and place and experiencing the world about you for the first time. Supermarkets feel too large. The city smells funny. Crowds overwhelm. Things that you once were able to filter out when you lived here suddenly shock and surprise you: twenty-four hour TV news shows, five-mile long traffic jams, polluted harbors, clouds of yellow smog, miles of concrete and steel buildings, mountains of trash.

For a brief time you have an opportunity to step back and feel yourself disengaged from all this madness. You feel scrubbed and cleansed, emptied of the impurities that you have lived with for so long. You feel as if you have undergone some kind of deep psychic conversion, and as with all such conversions, you feel as if there might be something important, some simple insight, perhaps, that you have learned along the way and that you might be able to share with those who have not been able to make the journey with you.

But even as you try to articulate your hard-won knowledge, you feel the moment slipping away. The sense of strangeness begins to pass, and you feel yourself growing comfortable again with all the trappings of a world that you have always accepted as normal. Then comes an even greater fear, not that you will remain a stranger, but that you will all too quickly begin to re-integrate, dressing your mind in the old familiar patterns, forgetting where you'd been and what you may have learned along the way.

Soon you start to grow anxious. You worry that one day you might find yourself strolling along a beach, listening not to the cry of gulls or the hiss of breaking waves against the sand but to sound of electronic voices from the headphones pressed against your ears. Or else you fear that one evening while you're driving home from work, you might find yourself drumming your fingers on the steering wheel of your car and staring impatiently ahead, fixated on the red traffic light that hangs above the roadway, totally unaware of the waxing crescent moon and the brilliant glow of Venus in the blackened sky beyond.

And so, inevitably, you start to plan and scheme, until you have devised a way to return once more to the Cabot Strait. You dream of crossing over, cradled by the swell, buffeted by the wind, washed by frigid seas, until you are thrust anew into contact with the natural world, cut off from the twenty-first century world of people you have so recently left behind, invited to wonder again about the planetary rhythms that give form and substance to this place and that define your relationship with it.

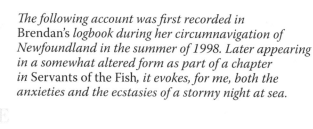

The following account was first recorded in Brendan's *logbook during her circumnavigation of Newfoundland in the summer of 1998. Later appearing in a somewhat altered form as part of a chapter in* Servants of the Fish, *it evokes, for me, both the anxieties and the ecstasies of a stormy night at sea.*

7
Gale Warning

THE SOUTHWEST WIND RATTLES IN the rigging, rolling *Brendan's Isle* against the pier where she is moored. This morning the sky is filled with mares' tails. I am awake before sunrise, seated at the navigator's station, poring over a stack of navigational charts. The feature on these charts that dominates my attention is a treacherous forty-mile section of coast that lies to the east and south of our present location: the infamous "Straight Shore" of Newfoundland.

The chart shows no safe harbors along this shore. The off-lying banks are dotted with low, reef-strewn islands. The waters close to the beach are shoal, filled with uncharted rocks. And just beyond Cape Freels, at the far eastern end of the strand, lies an area of shallow ledges and unpredictable currents so foul and uninviting that it has earned an equally foul and un-inviting name from local mariners: the Stinking Banks.

The winds for the past several days have been light. The latest weather forecast, however, calls for increasing westerlies, strengthening to south-west twenty-five to thirty knots by midday, then strengthening again and increasing to southwest gales thirty-five to forty by afternoon. Ordinarily, such a forecast would be ample cause for a little sailboat like ours to remain in harbor for the rest of the day, tucked up under the lee of the land and moored securely to a strong pier. On this particular morning, however, the forecast gales seem an advantage rather than a threat, for if the forecast-ers are correct, the wind will blow directly off the land, transforming the Straight Shore into a sheltered area of slight seas and fast, dry sailing for its entire forty-mile length.

Tucked up under the lee of the land and moored to a strong pier

As soon as breakfast is cleared away and the morning chores completed, my shipmates and I cast off *Brendan's* mooring lines and point her bows seaward, heading south and east toward the Straight Shore and the bays of southeastern Newfoundland. For the first six hours of this passage, the weather conditions develop almost exactly as the forecasters have predicted, and my decision to sail seems a good one. The day is clear, and the wind, gusting across the land, is warm. All this begins to change, however, as soon as the boat sails clear of Cape Freels and enters the area of shoal water that extends for nearly ten miles to the east.

Here, at the gaping mouth of Bonavista Bay, the wind accelerates along the shore, gusting well above forecast velocities. The air and water temperatures plummet as the boat draws away from the land, and soon a cold fog materializes, seemingly out of nowhere, obscuring the horizon and limiting the visibility to less than a mile.

Earlier in the day, while *Brendan* was still sailing under the lee of the Straight Shore, I had asked Amanda and Richard to help me reduce the rig to a fully-reefed mainsail and spitfire jib. But now even this small amount of sail begins to overpower the boat in the gusts that swirl up out of the gloom. I take over the helm as we draw abeam of the whistle buoy at Charge Rock, and I struggle to drive the boat a few more degrees to windward. Just over two miles ahead are the Stinking Banks. To the west and south of these is Bonavista Bay, with dozens of safe harbors at Valleyfield, Indian Bay, Deer Island, Lockers Bay, Sand Cove, and a maze of islands and thoroughfares beyond.

For an hour I battle with the helm—unwilling to admit that I have made a mistake in ignoring the morning's forecast. I try to force the sailboat into breaking seas and keep her head to wind—but she will not respond. She heels over, spilling the wind out of her sails, making more leeway than headway, and the Stinking Banks remain an obstacle that she cannot surmount.

"I get a feeling we might be spending the night out here," hollers Amanda, her voice cracked and distorted by the wind as the boat hobby-horses across crooked seas.

"Might be better than breaking the rig," cries Nat.

"Or breaking ourselves," I mutter, realizing that the only sensible alternative at this point is to place the boat in a defensive attitude and heave-to in the mouth of Bonavista Bay.

Often during a gale at sea, a sailor must learn how to move in concert with the forces arrayed against him. As the wind increases and the seas grow more chaotic, he must learn how to set aside his own agenda and surrender to the agenda that nature has prepared for him, seeking a safe compromise with the conditions at hand.

Some sailboats manage best under such conditions when they are brought head-to-wind, with a drogue or sea anchor fastened to the bows. Others ride most comfortably when they are left to "lie ahull," as sailors say, drifting on their own in the trough of the swell with all the sails removed. And still others are best served when they're allowed to run off,

Moving easily in a sheltered area of fast, dry sailing under the lee of the Straight Shore

powered by a tiny patch of sail to maintain steerage, dragging warps from astern, if necessary, to keep from running too fast down breaking seas and tripping over their own bows.

Once I make the decision to stop and place the boat in a defensive attitude, I have little doubt about which of these tactics will serve best, for *Brendan's Isle* is a boat that can ride to a following sea better than any I've ever sailed. I ask Nat to steer downwind and put the seas on the quarter while Amanda and I go forward to rig the storm trysail. Almost immediately the boat stops laboring. The spray that had been driving across the bows abruptly ceases. The noise of the wind drops. The crash of waves sluicing down the decks is replaced by the hissing sound of water flowing past the taffrail. Even without warps, the boat settles into a comfortable rhythm. With almost no effort on the helm, she moves off on an easy reach, slipping down the backs of seas, making her way slowly south and east.

The next twelve hours serve as a break in the progression of the journey—a hiatus in which there is nothing else to do but hunker down and wait for the gale to moderate. Beginning with Nat, each member of the crew agrees to take an hour's trick at the helm. Meanwhile, I remain below, on-call in case of emergency, but otherwise banished to the solitude of the cabins and the safety of my own bunk. For hour after hour I lie pinned against the leeboard, serenaded by the creak of the steering cable and the rush of water along the hull, confined to the company of my own thoughts.

For a time I entertain myself with a little game I'd devised during other nights at sea when the swell was just as large and the noise of the wind just as shrill. The object of the game is to somehow make sense of the cacophony of thumps and creaks and groans that the boat makes as she rolls across a deep seaway. First I create a mental inventory of all the sounds, ordering them from loudest to quietest. Then I attach an appropriate description to each sound—a wave sluicing down the deck, a cockpit scupper clearing its drain, a flogging jib, a squeaking bulkhead where an old glue-joint has failed, a glass bottle shifting position in one of the galley cupboards, a halyard slapping against the mast, a bilge pump cycling on and off. Each time I am successful in finding a satisfactory description, I remove the sound in question from the general inventory and store it away in a kind of imaginary recycle bin. In this way I'm able to eliminate sound after sound until—in my imagination at any rate—both the boat and the sea beyond become dead quiet.

On this particular evening, hove-to in the approaches to Bonavista Bay, I am able to find a name for every sound I hear and thus to put to rest whatever vague anxieties I might have brought below with me when I left the deck. Now I can close my eyes, breathe deeply, and let myself drift in and out of consciousness in a kind of semi-dream state, confident (at least until the advent of some jarring new sound) that my young shipmate at the helm has the sailboat under control.

The sound game, of course, hasn't always proven so successful. Fifteen years before this time during *Brendan's* first trans-Atlantic crossing, I had been far less able to cajole myself into sleep during my off-watch hours. For one thing, the boat was new and unfamiliar to those of us who sailed her—which meant that many of the sounds she made were still difficult to decipher. A clanking noise I'd never heard before would nearly drive me crazy until I'd risen from my bunk to search it out. Most often the sound would be benign—a flashlight rolling around in a drawer; a glass rattling in the sink. But just about the time I'd decide I could ignore it, it would prove to be the signal of a crisis in the making—a broken linkage on the steering vane, perhaps, or a loose drive wheel on the refrigerator's compressor pump.

Another source of worry on some of the early passages was that much of the rig, being new, was still untested under extreme conditions. Twice on that first Atlantic crossing I was awakened from a deep sleep to the sound of crashing noises on deck. The first time, six days out of the Chesapeake Bay on the southern edge of the Grand Bank, I awoke to an explosion that sounded like the report of a cannon, followed by a sudden deceleration in the boat's forward momentum and a muffled cry from the cockpit. The first thing I noticed as I scurried up the companionway steps was the torn head of the spinnaker flying at the masthead like the tattered banner of some defeated regiment. The rest of the sail had been shredded into several long streamers that were being dragged along the hull to leeward just beneath the surface of the water.

The second crashing noise occurred several hundred miles south of Iceland, during a sequence of gales that had stalked our little cutter for twelve uninterrupted days. *Brendan* was running that afternoon in near-gale conditions under a triple reefed mainsail and a yankee jib poled out to windward on an aluminum reaching pole. I was off watch, sleeping soundly, when the whole vessel began to jump in time to a series of concussions that sounded like someone trying to stove-in the foredeck with a pile driver.

Once again I found myself stumbling out of my bunk, pulling on boots and safety harness, groping in the semi-darkness toward the companionway steps. Up on deck the two watchkeepers seemed stunned, their jaws agape, as they stared at a broken section of the reaching pole spinning drunkenly from its attachment point at the clew of the yankee. Each time the sailboat accelerated down a wave, the pole—or what was left of it—made a complete rotation, driving its jagged end into the teak deck, sending chips of splintered wood into the air. One section of the pole hung limply against the mast, banging metal-on-metal with a rhythmic clank, while the other end spun like the blade of a wounded helicopter, defying anyone to go forward to try to wrestle it down to the deck.

Both emergencies—the torn sail and the shattered reaching pole—
were resolved without incident, and neither proved detrimental to the successful completion of the voyage. Each one taught a lesson, however, as
every emergency at sea is likely to do. The pole was strengthened with a
heavier extrusion and better end fittings as soon as we found a good machine shop in Europe. The spinnaker was mended by a well-recommended
Danish sailmaker, with a promise from the crew that from then on it would
be taken down whenever the wind started gusting beyond safe limits.

With these oddly comforting thoughts, I let my mind come back into the
present moment, hove-to and moving easily in the mouth of Bonavista Bay.
I roll over in my bunk once more, squeeze my eyes closed, pull my sleeping
bag up around my ears, and feel myself beginning to doze off, transported
into an almost euphoric sense of well-being by the now-familiar sounds of
our sturdy little vessel as she skids down the backs of following seas.

Shortly after midnight, the fog begins to lift and the wind veers, first into
the west, then into the northwest. Richard calls from the helmsman's station into the open window above my bunk, explaining that he has just spotted a light flashing on the horizon far off to starboard. I pull on my boots
and foul weather jacket and mount the steps into the cockpit. The wind
still whines in the rigging and the seas remain steep and unruly, but the air
has turned colder, and a gibbous moon has emerged from behind ragged
clouds to light the decks with an eerie silver glow. The flashing light on the
horizon, I know, is the lighthouse at Cape Bonavista, a bold headland that
marks the boundary between Bonavista Bay to the north and Trinity Bay
to the south. If we bear off, I explain to Richard, keeping the wind on our
quarter and the lighthouse on our starboard bow, we should close with the
cape by sunrise. Then, with luck, we might fetch up under its lee and find
safe harbor somewhere along the Trinity shore.

The night advances through the setting of the moon and the ponderous
rotation of the dipper around the northern star. Amanda follows Richard
at the helm. Liz follows Amanda. The dawn creeps into the eastern sky, and
the shape of the land grows on the horizon like a grizzled beast emerging
from the sea. The wind eases back as the sailboat doubles Cape Bonavista,
then eases again as she passes a series of small headlands at Spillar's Point,
Cape L'Argent, Flowers Point, North Head. Finally, as she draws abreast of
a fairway buoy at Poor Shoal, the high land to the north closes in behind
her and the wind drops in earnest.

To the east, a trio of humpback whales roll in lazy circles and blow in the pastel dawn. To the west a set of range lights mark the entrance to Catalina harbor, a landlocked bay that provides the most secure anchorage on this coast for a dozen miles in either direction. I fire up the diesel engine and follow a fishing boat into a buoyed channel that leads between a pair of rocky headlands and up to a small village. I hardly notice the men who stand clustered on the government pier or the rafts of fishing vessels moored on either side, so eager am I to conclude this long and difficult night. I simply guide the sailboat into an empty berth, make a cursory check of the mooring lines, and head below for a few delicious hours of deep and dreamless sleep.

Each minute of light experienced feels like one stolen from
a crushing winter. You walk gently about... with a sense
of how your body breaks the sunshine, creating shadows.
You converse in soft tones. The light is—perhaps there is
no other word for it—precious. You are careful around it.

Barry Lopez, *About this Life*

8 Northern Light

PREOCCUPIED WITH DREAMS OF PALM trees and white-sand beaches, most cruising sailors find the idea of voyaging into the far north to be simply unthinkable. For those who have been there, however, the reasons for sailing north are legion. One of the most compelling—one that beckons like a siren song—is the uncanny purity and persistence of the northern sunlight.

Just for the sake of providing a number, let's say that your perception of the change in the quality of the light begins as you sail across latitude 50 degrees north (although to be entirely honest, a degree or two on either side of 50 would probably serve just as well). You have been paying attention to other, more pressing matters—studying a set of sailing directions, conversing with your watchmate about an approaching vessel, changing down the head rig—and you simply haven't taken time to notice. But this morning as you look up from whatever task has been commanding your attention, you sense that something fundamental has changed. The line of demarcation between the surface of the sea and the jagged range of mountains to the east has become more distinct. The color of the water has deepened from olive drab to viridian green. The wave crests are whiter. The shadows cast across the deck by the sails and rigging appear knife-sharp, as if some invisible hand had removed a poorly focused lens from before your eyes.

The most dramatic change, however, comes later, as the sun drops closer to the horizon. Now the evening light begins to bathe the entire vista before you in a searing and unearthly whiteness. Unlike the evening light that you have grown accustomed to in the lower latitudes and that

As white as anything you will ever see in nature

illuminates the landscape just before sunset with a fiery glow, this light contains no hint of red. The sun, even as it drops the last few degrees toward the horizon, remains a blazing white orb—too bright to look directly into even for a moment. Shadows appear jet black as they stretch farther and farther into the east. The colors of the land grow luminous. A lighthouse on a nearby point of land seems to glow in an explosion of whiteness—so painfully white that it causes you to rub your eyes and look away, as if you had been staring at the sun itself.

During most of our northern voyaging aboard *Brendan's Isle*, the summer light has first become evident somewhere along the western coast of Newfoundland, as my shipmates and I have made our way northeastward on the 300-mile passage from Cape Breton Island toward the southern coast of Labrador. If the weather in a given year happens to be overcast, the change in the light may not become apparent until the sailboat has passed through the Strait of Belle Isle and entered an area of migrating icebergs and drifting sea ice that sailors refer to as "iceberg alley." Then, as the sky begins to brighten and the sun bursts though a hole in the cloud, a berg that has appeared all morning as an indistinct grey lump on the horizon will suddenly be transformed by a shaft of sunlight into a blazing white triangle of light, as white as anything you will ever see in nature. Once again, no matter how many times you may have witnessed this transition, the experience of the light will come as a shock and leave you, at least momentarily, awestruck.

For me, perhaps the most dramatic onset of the summer light occurred several years ago after a 1400-mile ocean passage, on the eve of *Brendan's Isle's* final approach to the Westmann Islands, near the southwest coast of Iceland. My shipmates and I had set sail ten days earlier from St. John's, Newfoundland, feeling our way in dense fog for 300 miles across iceberg alley and the northeastern Grand Bank, then racing for the next 1100 miles before a series of westerly gales across the top of the North Atlantic. As it does so often in this part of the world-ocean, the sky remained overcast for the entire 1400 miles. The sun shown only as a vague area of luminescence, slowly tracking across the sky behind a low, scuddy layer of cloud. The cloud itself streamed past the sailboat in long windrows, while the surface of the sea, without any shadow to provide definition, remained nearly featureless, marked only by crooked lines of spindrift and an occasional breaking crest that lifted under the sailboat's transom and skidded her downhill.

Sunrise at Cape Mugford, northern Labrador

The only noticeable change in the light for this entire passage was in the actual length of each day. As the sailboat climbed the latitudes, from 48 degrees north latitude when she left St. John's, through the 50s and into the low 60s as she approached southern Iceland, the length of time between sunrise and sunset grew inexorably longer. During the final days of the passage, the twilight also lengthened, until at last there was no longer any real darkness, but only an hour or two of soft afterglow on either side of midnight.

It was during this period of afterglow on the night of July 9 that *Brendan's Isle* found herself climbing up onto the Icelandic shelf, about 60 nautical miles southwest of the Westmann archipelago. The cloud cover had stopped moving overhead. The wind had grown light and fitful. Amanda stood at the helm, trying to keep the boat moving forward in the spiritless breeze, when she became aware of a brilliant crimson line, pencil-thin, burning like liquid fire all along the northern horizon.

Unwilling to trust the evidence of her own eyes, she stepped to the companionway and called me to the cockpit. She waited a few moments for my eyes to become adjusted to the strange scene before us. "Is that what I think it is, skipper?" she asked, inclining her head toward the shimmering line on the horizon. "If I didn't know any better, I'd say Iceland was on fire."

I sat down on the cockpit bench next to her, sniffing the air, gazing up at the overcast, checking the masthead pennant, feeling with the back of my neck for the tiny catspaws of breeze that roamed about on the surface of the swell. Nothing seemed unusual... except for that line of red light.

"It has to be the sunrise," I said at last. "It's just that the line between the overcast and the clear sky beyond is so straight. I've never seen a line of overcast just end like that, as if some master surgeon had sliced open the sky with a scalpel...."

For the next hour, as *Brendan's Isle* made her way in fits and starts toward her destination, the line of overcast rose higher and higher in the sky, eventually revealing an area of clear, cloudless sky beyond. As the line drew closer, the color in the clearing shifted from crimson to orange to robin's egg blue. The wind filled in from the northwest and soon began kicking up a sea of tumbling white crests. Just as the last of the overcast was passing overhead, the sun's upper limb broke the plane of the horizon with an explosion of white that illuminated the surface of the sea and lit up *Brendan's* sails as if they had been caught in a spotlight. Moments later, the loom of a 5000-foot-high mountain glacier, the Myrdalsjokull, appeared on the sailboat's bow, rising out of the sea like a distant plume of cloud. The glacier, whose glowing white presence would serve as a steering mark all day, was on the southern coast of Iceland, a full day's sail beyond our intended landfall and still more than 120 miles away.

Even as the clarity and color of the summer light are the qualities that most amaze you as you sail north, it is the *length* of the light (and the corollary absence of darkness) that actually begins to alter your body chemistry. The longer hours of daylight affect you in the same way that they might affect a flowering plant, feeding your body with warmth and energy, speeding up your metabolism, creating a kind of short-circuit in the sleep center of your brain. The result is a nearly continuous feeling of exhilaration in which you seem to need no rest.

The light begins to affect you just as soon as you turn your vessel north and start to climb the latitudes, for even a change of a few degrees at midsummer

The shadows cast across the deck appear knife-sharp

will translate into noticeably earlier mornings and later evenings, and these, in turn, will begin to make you feel as if somebody on board may have slipped a dose of amphetamines into your evening beverage. The farther north you sail, the more acute your response, until one day when you finally cross the Arctic circle and sail into the realm of the midnight sun, your body will simply incline itself like a flower toward the light, shudder with anticipation, and surrender to the endless pleasures of a world without night.

Brendan's Isle and her various crews and I have ventured three times into the land of the midnight sun. The first time was in 1994, the year we sailed to Disko Bay, Greenland, to roam among calving icebergs at the face of Kangia, the great River of Ice. The second time was in the summer of 2000, during our circumnavigation of Iceland, when we hove-to some thirty miles north of the Melvakkanes peninsula for the express purpose of raising a toast to Boreas Rex, ruler of the Frozen North, with a bottle of

The late evening sun lights the land at Nolsoy, Faroe Isles.

cheap Spanish wine and a frigid swim. The third time was in the spring and early summer of 2001, when we voyaged to northern Norway, some three hundred miles north of the Arctic Circle, there to spend nearly five weeks exploring the bays and thoroughfares of the fabled Lofoten islands and the northernmost capes of the Norwegian *skerregaard.*

This third (and longest) voyage to the Arctic began in southern Denmark, only a few weeks before the summer solstice. My three shipmates—Kell, Ben, and Bill—were all experienced sailors, all eager for the adventure of sailing north. None except Kell had ever experienced the northern light, however, and none, including myself, had ever spent such an extended period of time climbing the latitudes toward Arctic summer.

The effect that this experience had upon all of us was dramatic. Day after day, we seemed to grow more energetic, rising earlier, sailing later into the evening, remaining on deck to talk and socialize long after our watches were over, eating in irregular and sporadic bursts (or else forgetting to eat at all), finally climbing into our bunks at one or two o'clock in the morning, only to awaken a few hours later with the sunlight streaming into the cabins, feeling fully rested and ready to begin the new day.

I felt fortunate to witness the majesty and grandeur of this planet.

On the afternoon of June 6, near the end of the third week of this journey, the sailboat crossed the Arctic Circle in a remote area of mountainous
islands and deep, winding thoroughfares. Several miles north of the circle,
we made our way into a landlocked anchorage on the southwest side of the
island of Renga. Here, after a dinner of fresh-caught cod, my shipmates and
I decided to celebrate our arrival in the Arctic with a climb into the hills
above the harbor to view the surrounding landscapes and watch the sun as
it skirted along the horizon at midnight.

Once ashore, we tied the rubber boat to a boulder and set off to find
a suitable route for our assent. The terrain near the beach was steep and
rocky. In the gullies stood clusters of small birch trees, just beginning to
show their spring foliage. Higher up, the rock surfaces were smooth and
mostly barren, fringed with lichens and mosses and tiny Arctic wildflowers. We scampered single-file to a pinnacle outcrop about three hundred

feet above the harbor, where we stopped to rest and take photographs. After a short respite, my shipmates decided to continue on to the summit, another several hundred feet above. I chose to remain where I was, however, for I was in need of a few moments by myself.

After my shipmates had gone, I found a flat space to sit on the rocky outcrop. I gazed down at the sailboat lying alone at anchor, the only sign of human presence for as far as I could see. Beyond the boat, a hundred islands sparkled like a vast emerald blanket in the dropping light, their flanks casting long purple shadows across the channels that flowed around and among them. I stared in silence at the vista before me, overwhelmed by the sheer weight and density of its physical presence. Suddenly, without warning, I found myself crying, the tears streaming down my face until I could barely see. I tried to make some sort of utterance—a word of thanks, perhaps, to the gods, to lady luck, to the grand design (or randomness) of all this profusion of beauty—but the words would not come. Instead, as the tears collected warm and salty on my lips, I simply closed my eyes and let the moment sweep over me. I felt, in that instant, almost perfectly centered, as connected as one can ever be with the stuff of this world, and I understood that I was truly fortunate—to be alive, to be in this magnificent setting, to be sailing with friends aboard a strong little ship, to be afforded the chance to bear witness once again to the majesty and grandeur of this planet....

No matter how much I might have wanted to stop and savor that moment on the hillside at Renga, the moment soon passed. My three shipmates appeared on the rocky plateau behind me, and I stood and joined them for the climb back down to the beach and the ride in the Zodiac out to the sailboat. We celebrated the rest of that evening with a bottle of port and a box of Cuban cigars, forgetting about sleep until the sun had shown itself again through a notch in the hills, painting the sky a deep cerulean blue and flooding the decks with a light as clear and white as noon.

For the next five weeks the sun never set on *Brendan's Isle*. She sailed in the Arctic daylight for two more weeks with Kell and Bill and Ben, and then, after a scheduled crew change in the city of Tromso, for nearly three additional weeks with my wife Kay and several other friends. During all this time the sun tracked round and round the sailboat in a series of concentric circles, rising to its apogee in the southern sky at noon, falling to its perigee along the northern horizon at midnight. The intensity of the light on its diurnal circuit and the direction of the shadows were the only solar indicators of the days passing. The biorhythms of every living creature in this daylight world were temporarily placed on hold.

Finally one morning during the second week of July, I realized it was time to leave these latitudes and begin moving south. The summer was slipping past, and we still had fifteen hundred miles to sail on our return journey to southern Denmark. With a squally west wind on her quarter, *Brendan* set out across the Vestfiorden, bound for the little archipelago of Helligvaer. That night, anchored under the lee of low islands, we lingered after supper in the cockpit, watching the sun skid along the horizon, lighting the distant peaks of Lofoten in a riot of pastels and golds. As midnight approached, the sun's lower limb sliced into the sea, still moving left to right, but dropping inexorably until all that was left was a tiny sliver of light.

The next moment there was nothing. The sun had set, if only briefly, and *Brendan's Isle* lay in dusky twilight—a reminder of the darkness that would begin to overtake her now, night after night, until the Arctic light was just a memory and the gathering darkness a palpable fact, like the unexpected return of a prodigal child.

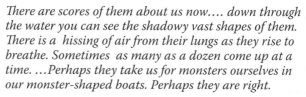

There are scores of them about us now.... down through the water you can see the shadowy vast shapes of them. There is a hissing of air from their lungs as they rise to breathe. Sometimes as many as a dozen come up at a time. ...Perhaps they take us for monsters ourselves in our monster-shaped boats. Perhaps they are right.

Frederick Buechner, *Brendan (A Novel)*

9 The Passing of Whales

FOR AS LONG AS HUMAN hunters have been sailing on the sea, the northern Atlantic Ocean has been both a breeding ground and a killing field for countless thousands of whales. The history of human predation on these largest of all terrestrial mammals has been erratic, rising and falling with the development of new tools and technologies, the changing sizes of whale populations, and various international treaties and agreements. Over the centuries certain species, such as the Atlantic gray whale, have been hunted to extinction. Others, such as the right whale, the bowhead, and the narwhal, have been reduced to remnant populations whose severely limited gene pools give them slim hope for survival. And still others, such as the minke whale and the long-finned North Atlantic pilot whale, have survived the slaughter with a surprising resiliency and continue to populate the northern ocean, if not in their original numbers, at least in large enough aggregations that they appear to be safe for now.

The ancient legends about the Irish cleric Brendan of Kerry tell of an age long before commercial whaling, when the northern ocean was still filled to overflowing with all manner of leviathans and creatures of the deep. According to the *Navigatio Sancti Brendani* (the widely popular medieval re-telling of the Brendan story), the Irish sailor-saint encountered whales many times during his seven-year odyssey. One of the most memorable of these occurred early in the voyage when he and his crew beached their curragh on a sleeping whale, mistaking the animal for an island, and then proceeded to light a cooking fire on its massive back. The whale, it seemed, found this experience rather ticklish and began to swim away, terrifying the crew and sending them into panicky retreat until Brendan assured them that the whale intended no harm. Brendan named the monster Jasconius, "the largest fish in the sea," and by the very naming, it seemed, he established a familiarity with the beast and bestowed upon it a kind of generalized beneficence. Every year afterwards for the next seven years he returned to the site of the encounter, there to meet and "swim" alongside Jasconius again.

In spite of the bloody slaughter inflicted upon the great whales in the intervening centuries, Jasconius and his kind continue to behave toward human beings as they had toward Brendan, with astonishing patience and restraint A dramatic case in point is recounted by Tim Severin, author of *The Brendan Voyage* and master of the twentieth-century replica of the boat that old Saint Brendan himself may have sailed to the New World. Like the vessel it was modeled after, Severin's leather curragh presented approximately the same silouhette as a medium-sized whale. Sheathed with slabs of oxhide and waterproofed with animal fat, it presented a shape and color and odor that might easily have been mistaken for a living animal.

Severin had studied the *Navigatio* in great detail and was well aware of the numerous whale encounters that his legendary predecessor had experienced. He was also aware of the animal shape and odor of his little boat, and he knew how simple it might be during his many weeks at sea for a curious whale to venture too close and damage or sink the fragile craft. Yet as Severin himself noted, if he had any hope of carrying out his plan of sailing a medieval skin boat all the way across the North Atlantic, he would just have to take his chances with the whales and trust that they would leave him and his strange little vessel alone.

Such, however, was not to be the case. Instead, from the very beginning of the voyage, the whales seemed fascinated by the leather boat. Day after day, whenever the weather was calm, the animals would approach, sometimes singly, sometimes in large groups, as if they were drawn to the vessel by some strange affinity. No one on board had ever seen whales behave like this—not even the Faroese crew member, Trondur, who had had a great deal of experience with whales and whale-catching.

Severin describes one of their early encounters:

> *Looking down into the water, we could see the extraordinary pattern of large moving shadows, as whale after whale moved gently under our leather hull, a vast congregation of animals changing places as they rose and fell, a living escort*

> *of sea creatures not more than six feet beneath the hull...*
> *"Grind! Grind!" cried Trondur, usually so phlegmatic, but*
> *now almost capering with delight. These were pilot whales,*
> *one of the smaller species, though the larger members of the*
> *school were half as long as Brendan and probably weighed*
> *as much as our vessel. They were unafraid, and moving very*
> *slowly. Trondur seized one of his slabs of whale blubber.*
> *"This is grind," he said. "Very good to eat."*

Severin admits he was relieved that Trondur did not have his whaling harpoon handy that morning, for as the North Atlantic pilot whale (*grind*) was a basic component of the traditional Faroese diet, this "phlegmatic" Faroe Islander might easily have tried to kill one of these gentle creatures. Instead, by the end of the voyage, every man on the little ship (including Trondur) had formed an almost mystical bond with the race of peaceful giants that accompanied them, like troupes of ceremonial bodyguards, across the top of the world.

For reasons that are not difficult to comprehend, the same species of small pelagic whale that most often visited Tim Severin's leather curragh (the long finned North Atlantic pilot whale or *Globicephala melaena*) is also the species that has most often visited *Brendan's Isle* and her various crews during our own meanderings around the northern seas. Long-finned pilot whales are a deep water species that still populate the North Atlantic in robust numbers. They are known to be highly gregarious. The description of their geographical range, from the mid-Atlantic coast of the United States to the Labrador Sea in the west, and from Morocco to Iceland, Faroe, and the Lofoten Islands in the east, might just as well be a description of the geographical range of *Brendan's Isle* as she has criss-crossed the North Atlantic basin over the past two and a half decades.

The first time my shipmates and I were visited by a family group of long-finned pilot whales was in the midst of a gale, two hundred miles south of Cape Farwell, Greenland, during our first Atlantic crossing. The sailboat was moving at hull speed, accelerating down the faces of twenty-foot following seas, with sheets of spray flying off her bows and her hull and rudder vibrating with a high-pitched whine. Andrew stood alone in the cockpit clutching the big teak-and-holly steering wheel, swaddled in layers of heavy clothing and sheathed head to foot in a suit of white vinyl foul-weather gear. At one moment, all he could see when he glanced astern was the dull green face of a huge following sea, its surface streaked with spindrift and its top capped with a tumbling white crest. Then in the next moment he looked again to see the same wave filled with glistening black bodies, half a dozen on one side of the sailboat, half a dozen on the other.

As the boat skidded down into a trough and the crest of the wave moved ahead, the whales sounded and seemed for a moment to have disappeared. Then as the next sea mounted under the transom, the animals rose again, their pot-shaped heads and blow holes and huge black dorsal fins emerging in unison from the surface of the wave, moving without apparent effort at almost exactly the same speed as the sailboat.

Andrew stood motionless at the helm and stared in amazement at the surfing whales. The largest were perhaps twenty feet long and four or five tons in weight. Each time they rose to the surface, there was a hissing sound of a dozen short exhales, followed by the sucking noise of a dozen quick inhales. The whales nearest to the sailboat slalomed back and forth, sometimes approaching to within a body-length of the green fiberglas hull before sliding away again.

Andrew watched with growing concern as the whales drew closer and closer to the boat. He had been hesitant up to this point to call for help from below, for he knew that supper had been served only a few minutes before and that in these rough conditions his shipmates would be performing something of a balancing act as they tried to keep bowls and dishes and cups from sliding off the dinette table and crashing onto the cabin floor. Next moment, however, he found himself thumping his rubber sea boots on the cockpit grate and calling at the top of his voice for all hands.

Even as he called, one of the whales separated itself from the main group and drew alongside the sailboat, swimming close enough now that, if he'd wanted to, Andrew could literally have reached out and touched the thick black tip of its dorsal fin. Then, just as the faces of several bewildered shipmates appeared at the companionway, the whale moved forward, tucked itself under the flare of the bow, and began to ride the eddy under the cutwater like a playful dolphin, its great black flanks only millimeters away from the fiberglas hull.

Over the course of the next twenty-five years, *Brendan's Isle* and her various crews and I have witnessed dozens of variations on this close encounter with the pilot whales. Sometimes the experience has been similar to the first one, with a cold wind keening in the rigging and mountainous seas tumbling pell mell across the sailboat's wake. Other times the surface of the sea has been glassy smooth, with the air so still that even before anyone could see the whales approaching, we could hear the clicking and chirping of their whalesong through the thin walls of the sailboat's hull.

One day in 1994 while *Brendan's Isle* was crossing the Davis Strait on her way to western Greenland, we were treated to several hours of the most exuberant whalesong I have ever heard. The gray sky and sea were as one that morning, so close in color that the line between them was barely distinguishable. A long, undulating swell moved noiselessly under the hull, with just enough air moving across its surface to propel the sailboat in slow motion toward the east. The whalesong, barely audible at first, grew more and more persistent, slowly increasing in volume until the entire vessel resonated with its strange, inhuman music.

The whales that eventually appeared on the southern horizon were not an ordinary family group of ten or fifteen individuals, as we had so often encountered in the past, but were instead a massive aggregation of hundreds of whales—males, females, juveniles, newborns—all moving together in a perfectly choreographed dance. As they drew closer, the entire assembly turned and began to follow the sailboat. A few of the larger males moved out ahead of the herd and closer to the boat, although none seemed willing to approach nearer than a few hundred yards. The cabins by this time reverberated with a cacophonous jumble of whistles, clicks, keens, groans, high-pitched coos and moans.

The whales were communicating intelligently with one another—this much was already clear from the high degree of organization we were able to observe in the herd. Was it possible, I wondered, that these highly evolved, large-brained mammals were also trying to make intelligent contact with the clumsy green sea creature they were following?

Feeling a little like a character in a grade B science fiction movie (but determined to follow my hunch), I stepped below into the main cabin and began flipping through the sailboat's collection of classical music CD's. If we were able to hear the whales through the sailboat's hull, I reasoned, then surely the whales should also be able to hear us. All I needed was to find a piece of music that would appeal to their cetacean sensibilities.

Let's see...Vivaldi, Brahms, Albinoni, Mahler, Hayden, Mussorgsky, Bach... Ah yes, J. S. Bach... a composer whose music is metronomic, repetitive, almost mathematical in its precision—just the sort of music, I thought, that might be useful in making contact with a race of intelligent aliens. I decided I'd try them first with one of Bach's Brandenburg Concertos.

No sooner had I placed the disk on the CD-player and hit the "play" button than I heard Amanda's frantic voice from the cockpit. "What are you doing down there?" she cried. "The pilot whales... they've stopped following us...I think they may be turning away...."

Without a moment's hesitation, I removed Bach from the CD-player and began searching for something different. I decided this time to look for a piece that was soulful, rhapsodic, full of power and raw emotion. Ah, I thought, here was something that might do—Aaron Copeland's "Fanfare for the Common Man"—an impassioned, melodramatic piece with lots of cymbals and trumpets and kettle drums.

Moments after I'd pressed the start button for the "Fanfare," I heard Amanda's voice again from the cockpit. This time, however, her words came in a kind of plaintive half-whisper. "Mike, come up here... you have to look at them...oh my god you have to see...."

I scampered up the companionway steps. To the right, in the direction that Amanda was pointing, a large male had approached to within a few feet of the sailboat's transom. Just behind him swam a pair of smaller females, and next to each of them, drafting in the eddy behind each one's dorsal fin, swam a tiny newborn, each no more than a few feet in length.

As the "Fanfare" grew in intensity, more and more whales approached the sailboat. Soon there were black bodies everywhere—underneath the keel, along both sides of the hull, under the bows, beneath the transom, alongside the rudder—so close they seemed to be caressing every contour of the boat's underbody. Punctuating the sound of Copeland's music were the sounds of the whales' breathing—the high, quick, pizzicato clicks of the infants and juveniles—the lower, guttural whistles of the adults—like the popping and wheezing of some great baroque pipe organ.

In the years since our first encounter with the pilots, I had read everything I could find in the scientific literature about them, and nowhere had I seen descriptions of the sorts of behaviors we were now witnessing. I had read about their gregariousness and their close-knit family relationships. I'd read about their tendency to congregate and feed along the edges of continental shelves. I'd read about their pattern of gathering from time to time into large migratory herds for the purpose of moving from one seasonal feeding ground to another. I'd read about their large brains (larger per unit of body weight than any other living creature ex-

A family group of long-finned pilot whales races to overtake the sailboat.

cept human beings) and about their powerful, highly developed echo-locating sonar.

But nowhere (except in the pages of Tim Severin's *The Brendan Voyage)* had I read that groups of these whales might approach a small vessel, play under its hull, draft in its bow wave, cavort in its slip stream. (In fact, according to several well known field guides, such actions were specifically excluded from the pilot whales' behavior.)

In the end, all we could do was to marvel at the whales' presence and enjoy their playful antics. Although I never played Bach for them again, I experimented with other short selections from Copeland, with parts of Brahms' First and Second Symphonies, with a few of Dvorak's Slavonic Dances, and finally with a long selection from Shostakovich's Symphony from the New World. They seemed to grow quickly bored with Brahms and Dvorak and appeared rather lukewarm toward the Russian. But every time I returned to Copeland they seemed to perk up, and as soon as I played the "Fanfare" again, they became instantly agitated, gathering around the sailboat in ever larger numbers, passing around and under her ever more closely, as if they were preparing to lift her on their backs and carry her off to some cetacean never-never land.

During the years following this encounter, the dozens of meetings that we have enjoyed with pilot whales have been so moving and so intense that I am tempted to describe them collectively as a kind of love affair—if not between the whales and the people, then certainly between the whales and the clumsy green "creature" that wallowed upon the surface of the sea and sang to them with trumpets and cymbals and crashing kettle drums. Every fresh encounter has been an occasion for my shipmates and me to break out the Copeland and watch the whales respond—every one, that is, until an encounter that began early one July morning several years ago, as *Brendan's Isle* was approaching landfall on a passage from Iceland to the Faroe Islands.

Once again, as so often in the past, a large family group of pilot whales appeared out of the haze that morning and began to approach the sailboat. This time, however, I issued a rather upsetting set of instructions to my crew. We wouldn't play Copeland's "Fanfare" on the boat's CD player—we wouldn't play music of any kind. Instead, we would roll in the jibs and fire up the diesel engine, revving it as fast and loud as it would go. If this noise didn't convince the whales to turn and swim away, we would find wooden blocks, shoes, boots, books, anything that was heavy enough to beat against the decks and the coach roof and the sides of the hull. We would yell profanities, stomp our feet on the cockpit grate, and in general make such an awful racket that the devil himself would want to turn and swim away.

The reason was simple. We were about to make landfall that morning in one of the few places left in the world where the hunting and killing of pilot whales was considered a birthright by a vast majority of the local population. Furthermore, we were arriving at a time of year when the whale kill (Faroese: *grindadráp*) was in high season, a time when as many as a thousand whales were about to be driven into cul-de-sac fiords all around these islands and slaughtered execution-style, until the waters of the fiords were stained dark red with their blood.

The first time *Brendan's Isle* had sailed to these islands, sixteen years earlier, neither I nor any of my crew had known about the traditional Faroese whale hunt. The first night we'd arrived, in fact, we'd been offered a

With the aid of Copeland's "Fanfare," a family group of pilot whales draws closer to the sailboat.

plate of whale blubber by one of the local citizens as a kind of welcoming gift, and we had done our best to appear appreciative. The next day we'd discovered portions of whale meat and blubber packaged in styrofoam trays and wrapped in clear plastic cling-wrap in the meat section of the local grocery store.

We learned about the details of the *grindadrap* not from the Faroese themselves (who tended even then to be rather secretive about the subject), but from a young German sailing couple, Kris and Edda, who happened to be moored alongside us for several days in their little homebuilt cutter *Anya Syra*. Kris was a photographer for a German nature magazine; Edda was a writer. They had been anchored the week before in a remote fiord in the northern end of the archipelago, photographing and collecting field notes for an article on nesting seabirds, when they inadvertently became eye-witnesses to an ancient blood-ritual: the hunting and driving and killing of almost three hundred long-finned pilot whales.

Edda was the one who provided the most graphic description of what they had seen. In the early stages of the hunt, she explained, a large migrating herd of whales was driven into the mouth of the fiord by dozens of small open boats. The operators of the boats, communicating with handheld VHF radios, swerved back and forth in snakelike patterns, always keeping the whales ahead of them, while their crews beat against the sides of the boats with long sticks and tossed weighted ropes at the whales to frighten them and keep them moving forward.

Slowly, relentlessly, the jaws of the trap were closed. The noise of the sticks drumming against the hulls and the roaring of the outboard engines echoed loudly in the bowl of rock that formed the closed end of the fiord. At the head of navigation, a large group of people gathered in the center of a semi-circular beach, while ahead of them another group stood in the cold water, armed with gaff hooks and long steel knives.

As the whales were driven closer and closer to the beach, Edda could see that they were hyperventilating. Everywhere their breath came in quick, short gasps. They circled and twisted and rolled. Whenever they tried to dive, they came quickly to the surface again with sharp sucking noises. The question Edda had been asking herself all morning, about why the whales had not simply descended into the deep waters in the center of the fiord and escaped back to the sea, now found its answer. It was no wonder the whales didn't dive... for the fact was they *couldn't* dive. They had been whipped into such a frenzy of fear by their pursuers that they could no longer control the complex process of oxygenation that had come so easily to them in the freedom of the open ocean. Their fear made them prisoners of the air, and the hunters understood, if only they could keep their victims confused and panic stricken, that there would be no escape.

From a distance, the slaughter of the whales might have appeared almost orderly, but from where *Anya Syra* lay anchored, near the western end of the beach, there seemed no order but only incomprehensible chaos. An engine racing out of control. An iron gaff hook swinging in a slow arc at the end of a rope tether, curving in slow motion toward the water and slicing into black flesh. Men crying out to each other in guttural syllables. A boat towing one gaffed whale at the end of a rope, riding up on the back of another, the blades of the outboard engine slicing a pattern of razor-thin wounds into the flesh. A black head rising vertically out of the water, its tiny eyes darting about the horizon, searching for a means of escape. The short, frantic inhalations of a female whale, in the act of giving birth, pinned with a harpoon against the side of a boat, whipping the blood-filled water with the flukes of her tail. Seabirds circling, screeching, diving above the death throes of a drowning bull. An infant whale floating upside-down, its distended belly bleeding from a series of small, jagged wounds, while others nudged it from underneath, in a futile attempt to roll its blow hole into the air.

Brendan's Isle *rafted alongside Kris and Edda's little steel cutter* Anya Syra *in Torshavn, Faroe Isles.*

On the shore, the body count steadily mounted. By mid-morning the beach looked like a battlefield, with row after row of mutilated black carcasses drying in the sun. A group of women and boys hauled the glistening bodies of the newly executed whales onto the upper part of the beach with a block and tackle, while others stood in the shallows, inserting grappling hooks into the blow holes of living whales, readying them for slaughter. Wading among the tethered animals, a line of executioners stood shoulder-deep in the blood-stained water, slashing with their knives at the necks and spinal columns and of the animals that still moved among them.

Edda and Kris turned away from the beach, unable to stomach this business any longer, and began preparing their boat for sea. Working in silence, they moved to the foredeck and began coiling down the anchor chain. After the anchor was stowed, Kris stepped to the mast and hauled on the mainsail halyard. Edda moved to the tiller and, with a sudden flood of tears, pointed their little cutter toward the open mouth of the fiord.

For twenty years, ever since the day I first listened to this description of the *grindadrap,* I've been trying to come to terms in my own mind with the violence and brutality of this event. The simplest way would seem to be to deny one's own involvement and place the blame entirely on the Faroese hunters, imagining them (as we often do with an enemy in wartime) as heartless, bloodthirsty savages, fundamentally different from other human beings. Indeed, these "strong young men" (*"raske drenge"*) might even be seen to incriminate themselves as they boast, in the words of their own battle song, "to kill the grind, that is our pleasure!" *("grind at droebe det er vor lyst")*.

For me, however, this attempt to bestialize the Faroese whale-hunters has never quite worked. For one thing, I have met and befriended many of the people who have taken part in the slaughter, and I know them as kind and pleasant human beings, not all that different from you and me. For another thing, I know that the vast majority of us—myself included— participate in the slaughter of the animals we eat—cattle and sheep and pigs and chickens. True, most of us do not take active part in the killings. Instead, we engage a group of proxy killers to perform the bloody deeds for us, enabling us to live in permanent state of denial over our own roles as a members of nature's most powerful predator species.

How, then, should we think about the killing of the pilot whales? Perhaps the best way might be to take our cue from the whales themselves, for with their innocence and their sacrifice, they point the way toward an answer. As they are driven against their wills to the slaughter, they ask us to look without flinching at the killing fields—those of us who can bear the sight—and to admit of our own complicity in the acts that are committed there. Then, after we are convicted by the evidence of our own eyes, they forgive us, with a magnanimity that is almost beyond our human capacity to understand.

Everywhere in the wide Atlantic the survivors return to us—without anger—without fear—without the desire for revenge—and they grace us once again with their presence. They swim alongside our boats, they draft in our bow waves, they sing along with our music, and they provide us with a model—if only we can bring ourselves to comprehend—of how we might rise above our own base nature and learn to live (and die) with compassion and dignity.

Rule One: Carry out everything that you carry in.
David Brower, *The Sierra Club Wilderness Handbook*

10 Footprints

ONE DEAD-STILL AFTERNOON IN EARLY August, the 18-inch propeller on the end of *Brendan's Isle's* stainless-steel drive shaft suddenly stopped turning. The sailboat, which had been proceeding under auxiliary power down the coast of southern Labrador, slowed and drifted to a standstill. Beth pulled back the throttle and eased the gear lever into reverse. No response. She tried the gear in forward again. Still no response. Although the full extent of our dilemma wouldn't be evident for another twenty-four hours, it was already clear that our vessel had sustained serious damage to her transmission and drive train. Now, on a remote northern coast some sixteen hundred miles from home, and without access to a mechanic or machine shop of any kind, *Brendan's Isle* drifted on a glassy sea, unable to proceed on her homeward journey until we could fashion a repair.

A day later the verdict was in: a leaking gasket and a slow loss of transmission fluid had eventually caused every moving part in the gearbox to be seized together and ground into shrapnel. The boat would have no mechanical propulsion until a new transmission unit could be shipped from the supplier in Chicago to a remote outport in southern Labrador, a process that could easily take several weeks.

Oddly, none of the five young sail-trainees on board seemed the least demoralized by this turn of events. As Beth explained to me the day after the breakdown, she and her shipmates had come to Labrador to contact the natural world and explore a wilderness coast. What did they care how fast or how far the boat traveled? Perhaps, while we waited for our new part, we could slow down—*way* down—and begin investigating the boat's surroundings entirely under sail, learning to pay attention to the details of this landscape in a way that we had been unable (or unwilling) to do when we had the option of using our engine.

The days that followed were filled with serendipity. The wind remained light, so the sailboat seldom traveled more than a few miles between anchorages. On several days she never traveled at all. The added hours at anchor provided time for forays ashore for photographing and beachcombing and collecting samples of the Arctic flora. Evenings in the cabins were spent talking quietly or reading by lamplight (in order to conserve our limited battery power). Dishes were washed in cold saltwater, delivered to the galley sink by hand pump. Day and night, the cabins were filled with the soft hissing of the catalytic diesel heater.

Slowly, inevitably, a silence overtook the little sailboat, mimicking the silence of the surrounding terrain. On certain mornings, when the frost lingered on the spruce and tamarack on the shore, we would lie for an extra half hour swaddled in our sleeping bags, watching our breath condensing on the portholes next to our bunks, listening to the wavelets lapping against the boottop and the barnacles cracking on the sailboat's hull. Then, as people began to dress and move about the cabins, our first words would be in whispers, as if we were in a holy place and did not want to break the silence too quickly with our voices.

A week after the breakdown I received a radio message that our replacement transmission would be arriving by floatplane at a nearby village in two days. This gave us one more night to linger in one of the wild, uninhabited anchorages in the area before having to return to the task of repairing the engine and continuing our homeward journey. With settled weather and a gentle west wind, we were able to sail that morning to a small archipelago, known collectively as the American Islands, and to anchor in a narrow, landlocked channel surrounded by barren hills.

The mood of the crew was giddy, almost intoxicated, as we set a large fisherman anchor in mid-channel and tied the stern to a boulder on the shore. As soon as these tasks were completed, Kirk and Scott announced that they were going to explore the seaward side of the islands, where they'd seen several small icebergs stranded along the beach. Beth and Tim decided they would climb to a nearby meadow to search for bake apples and partridge berries. Amanda, meanwhile, indicated that she was going to commandeer the Zodiac in order to make soundings of the anchorage for a sketch chart for the sailboat's log book. Each plan was articulated with a kind of breathless excitement, with everyone trying to talk at once,

Several crew members explore the hills at the American Islands.

as if this were the final afternoon and these islands were the last stopping place of the summer's journey.

Several hours later, with all the various agendas accomplished, the group found itself back aboard *Brendan's Isle*, gathered around the cockpit. This time Tim spoke for all the others. "We've taken a vote," he announced. "Before dinner we're going to climb to the top of the tallest island and build a cairn with all our names on it. It will be a way of marking our visit here for anybody else who comes after us."

With this announcement he pulled a small American flag out of his pocket—one of the dozen or so that I had stowed aboard *Brendan's Isle* that summer to give away to Inuit children as gifts—and he passed it around the cockpit along with an indelible black magic marker. "Here," he said. "Each of you sign your name on one of the stripes. Then we'll fasten it to the top of the cairn…an American flag for the American Islands…get it?"

Once again the mood of my young shipmates was giddy and excitable as they passed the little cloth banner from one person to the next. When it finally came around to me, I took the pen in one hand and the flag in the other. Then I paused, uncertain whether I really wanted to participate in their childish game.

"How about it, skipper?" asked Tim, with a distinct note of challenge in his voice. "Are you with us or against us?" Glancing about the cockpit, I saw the same question inscribed on every face. I hesitated only a few seconds longer before succumbing to the pressure of the group and making my mark along the bottom of the flag.

Tim was first into the Zodiac, first onto the beach, first to scamper up the steep hillside toward the summit of the tallest island. The others followed single-file behind him, while I took up the rear. By the time I had found my way past the final obstacles of the ascent and had reached the summit, my shipmates were already hard at work collecting rocks of various sizes and fitting them together into a small circular tower. Tim had gathered a stockpile of pebbles for use as a kind of "dry mortar" to hold the larger rocks together as well as to fashion a mount for the desecrated American flag that would fly atop the whole affair.

As the structure neared completion, Beth withdrew a small Nikon camera from the pocket of her parka and began snapping photographs. Several of her shipmates arranged themselves in heroic poses around the cairn, as if they had just conquered Everest and were in the act of installing the banner of their victory at the summit. Somewhat reluctantly, I joined several of the photographs. But as my shipmates continued their celebration, I found my attention wandering—out past the little mound of rocks that we had built to the majestic vista of the surrounding terrain, to the dark forms of islands receding one behind the other, the searing whiteness of icebergs grounded on their shores, the flat surface of the sea glistening like a mirror in the late afternoon sun. When I glanced back at the cairn, I suddenly understood that without intending, I had helped to desecrate more than just a miniature American flag. And I felt ashamed.

Later that evening, after sharing supper in the main saloon with my shipmates, I retired to the cockpit to watch the sun set over the hills of mainland Labrador. As the light dropped and the long, purple shadows advanced across the anchorage where we lay, I let my eyes wander upward to the tiny form of the cairn on the hill behind us, and I forced myself to reflect on the events of that afternoon.

One part of me wanted to rationalize our motives and somehow justify the actions that we had collectively performed. After all, I reasoned, the aboriginal inhabitants of the far north had been building cairns since time immemorial—cairns for marking everything from walking trails and sailing routes to food caches and favorite camping sites. European visitors had followed with ever larger and more dramatic markers to announce their arrival and passage across these same northern lands. I had recently read about a group of Viking monuments found in northern Greenland that bore the runic inscription: "Erling Sifhvatsson and Bjarni Thjordarson and Eindridi Jonsson built these cairns." After the Vikings came the Basques, the English, the French, the Irish, the Spanish, the Portuguese, all adding their stony signatures to the far-northern skyline. Surely, I told myself, if our European forebears had been free to leave such marks upon this land without fear of censure, why shouldn't my crew and I be allowed to do the same?

For a few moments such rationalization served to mollify the nagging sense of guilt I was feeling—until I found myself thinking about other, far more onerous marks that these same forbears had collectively impressed upon this land—marks that had altered it permanently, yet that remained virtually invisible to present-day visitors.

The little archipelago where we were anchored served as a dramatic case in point. In the centuries before the arrival of the Europeans, these islands would have been home to countless thousands of nesting seabirds: great auks and gannets, murres and puffins, Eskimo curlews and laughing gulls, cormorants and lesser auks, kittiwakes and Arctic terns. During the winter months, when the islands were sutured to the mainland by shorefast ice, they would have become a foraging ground for wandering family groups of white bears. In the spring, the beaches would have been blackened by massive runs of caplin, while the surrounding waters would have become home to large populations of walruses and harp seals. In the summer the same waters would have hosted pod after pod of pilot whales and humpbacks, orcas and minke whales, feeding to their hearts' content on huge clouds of krill and shoals of fat northern cod.

That evening, as I sat in the gathering dusk in *Brendan's* cockpit, I realized that, with the exception of certain of the whales, none of these creatures was to be found in southern Labrador any longer. Except for remnant populations on remote offshore islands, every species of bird I had thought about was gone now from this place, victims of centuries of exploitation by egg and feather collectors as well as by fishermen who used their flesh for fish bait. Some, such as the great auk and the Eskimo curlew, were gone forever from the Earth.

The white bear, whose territory once extended into boreal regions far to the south, was now so rare in every region but the high Arctic that its name had been changed by those who still hunted it to "polar bear," and its former range and robust numbers had been all but obliterated from our collective cultural memory.

The walrus population, once in excess of a million individuals across eastern Canada, had been reduced by the final decades of the nineteenth century to a few small breeding colonies in northern Labrador and the Canadian archipelago. The vast herds of harp seals had been hunted so relentlessly and with such chilling success in the Gulf of St. Lawrence and along the Labrador coast that the survivors were now on the order of a tenth or less of their original numbers.

Meanwhile the local population of humpback whales, a species that had once made its summer pilgrimage to eastern Newfoundland and southern Labrador by the thousands, had been reduced by the late 1960s to a remnant population of several hundred individuals. For the next three decades the few whales that managed to return found themselves in constant danger of becoming entangled in cod traps and either suffocating or being shot by fishermen trying to save their gear. Ironically, it took the catastrophic collapse of the cod fishery in the late 1980s and early 1990s (and the subsequent removal of the traps) to save the whales. In the ensuing decades, while the northern cod fishery ended and the cod itself all but disappeared from east Canadian waters, the humpback population actually experienced a modest rebound.

Sadly, the slaughter and disappearance of so many indigenous species was only part of the story of the degradation that had occurred along this coast since our forbears had arrived here. The other part of the story was more recent, occurring as it had as a result of the impacts of modern industrialism.

A glance at the stunted forests almost anywhere in inland areas of eastern Labrador immediately revealed the ravages of more than a century of acidification of lakes, streams, and surface soils by sulphuric acid (SO_x), nitric acid (NO_x) and other airborne pollutants transported here in the form of "acid rain" from the industrial centers of the United States and Canada. And although specific connections remain poorly understood, such acidification (and subsequent weakening of various forest species) had almost certainly contributed to the virulence of woodland epidemics such as spruce budworm, an insect-born disease that had periodically denuded and eventually killed millions of acres of black spruce, balsam fir, tamarack, and pine throughout eastern Canada.

As regards human disease, the aboriginal populations here had been decimated time and again by outbreaks of plague, smallpox, cholera, and viral influenzas brought to the North American continent over the centuries by European settlers. In more recent times, the chemical analysis of breast milk from Inuit mothers had begun to reveal another silent killer: a dangerously high level of industrial PCBs that had been entering the food chain from sources thousands of miles to the south and west and that was presently contaminating the flesh of animals and fish throughout the Canadian north.

Meanwhile, climatological measurements spoke of an even greater threat. Warming winter air temperatures, accelerating sea level rise, reduction in the thickness and extent of winter sea cover, retreat of the Greenland ice cap, freshening of the surface layer of the Labrador Sea and subsequent changes in patterns of deep convection, acidification of the ocean surface by atmospheric CO_2, all of these new and onerous shifts in Arctic climate patterns pointed toward the emerging danger (and as yet undetermined consequences) of human-generated climate change.

It has always been easy for a first-time visitor to this place to imagine it a pristine wilderness, unaltered by the tampering of human beings. Yet ever since some nameless European bludgeoned the first great auk or harpooned the first right whale or trapped the first northern martin or shot the first white bear or raided the eggs from the first Eskimo curlew's

My crew and I built a cairn at American Islands, Labrador.

nest or jigged the first northern cod, the impacts of our collective presence have been multiplying upon this land. Finally, if the visitor is honest with himself, he must begin to think upon these things. And then the tiny cairn that a group of careless sailors built upon the hilltop will become symbolic of something larger and far more dangerous... a symbol both of our past transgressions and of the great number of un-self-inspected alterations that we continue to impress upon this fragile land.

Twelve hours after they had finished building the cairn, my young ship-mates gathered for breakfast around the dinette table in *Brendan's* main saloon. In spite of the fine weather that morning and the promise of good wind, their collective mood was oddly subdued. Nobody was bragging about the "conquest" of the American Islands, as they had been the eve-ning before. Nobody was making jokes about the desecrated flag that still fluttered from its perch atop the nearby hill. Even Tim, staring down at the table and crunching loudly at the contents of his cereal bowl, seemed uncharacteristically pensive.

Beth was the one who finally posed the question. "Do we have half an hour, skipper, before we need to get underway this morning?"

"Sure, " I said, "but what's the…?"

"We've been thinking—all of us—and we've decided something needs to be done about that cairn before we leave this place. I promise it won't take very long."

Even as I nodded my assent, Tim was clambering up to the cockpit, untying the Zodiac from its customary cleat on the stern rail, and pulling it alongside. Moments later, with his shipmates all on board, he shoved off and pointed the bow toward the same place on the shore where they and I had landed the previous evening.

I watched from Brendan's cockpit as they disembarked from the rubber boat and made their slow assent toward the summit. I still was not certain what all this commotion was about. Had some of the stones in the cairn fallen during the night and were they embarked on a repair mission of some sort? Had they decided the tower was too small to be seen by approaching vessels, and were they going to enlarge the size of the structure? Or were they unhappy, perhaps, with the way they had fastened the flag to the top, afraid it might topple in the wind without further reinforcement?

The answer was not long in coming. As I watched, I saw the little group gather in a circle around the site of the tower, as if they were about to enact a sacred rite of some sort. Amanda made the first move, plucking the flag from the top, rolling it up on its wooden flag stick, and inserting it into the pocket of her shorts. Scott then squatted down, picked up a stone from the top of the structure, and tossed it ceremoniously over his shoulder. Each of the others followed in turn: Tim, Beth, Kirk, Amanda, and back to Scott again. Around and around the circle they went, each taking a stone and tossing it behind them until all the stones were scattered on the ground— just as they'd been twelve hours earlier—and the marker that was once intended to memorialize our passing here was itself only a memory.

I was overtaken by a jumble of emotions as I watched the last of the little tower disappear. I was encouraged… surprised… gratified by the action my shipmates had just taken. And I was proud… I don't think I've ever been as proud of anyone I've sailed with as I was of those five young people that morning.

I knew of course that the gesture they had made was nothing more than that…a gesture. On one level, it meant nothing at all. For neither the building of the cairn nor its subsequent destruction had done anything to alter the history of this place, to eradicate the degradation that had occurred over the centuries, to reset the clock and reverse the direction of humankind's collective impacts here.

On another level, however, their methodical dismantling of the cairn was a statement, no matter how small or inconsequential, that contained at least a glimmer of hope—hope that somehow our species might have the capacity to change—hope that a new generation might learn to come to a place like this and to perceive it not as a land to be exploited and manipulated but as a fragile wilderness, to be savored and protected from further harm.

Only time would tell whether the gesture these five young sailors made that morning would point toward something beyond itself, some future action they might perform, individually or collectively, that might have an impact on a larger scale. In the meantime, they had fashioned a simple statement, expressing their respect for this place by removing, as best they could, the footprints they had made in their passing.

The new Vale of Tempe may be a gaunt waste in Thule...a place perfectly accordant with man's nature—neither ghastly, hateful, nor ugly; neither commonplace, unmeaning, nor tame; but like man, slighted and enduring; and withal, singularly colossal and mysterious in its swarthy monotony.

Thomas Hardy, *The Return of the Native*

11   A Gaunt Waste in Thule

F OR MANY WHO HAVE LIVED in particular settings for a significant portion of their lives, there are landscapes that exert a kind of magnetic attraction and that make them feel safe and comfortable—landscapes that feel like "home." For some, this attraction is so strong as to seem like a genetic imprint, passed from generation to generation, the same way a salmon imprints on the particular river that was his birthplace.

My mother feels this way about the flatlands of central Michigan. She lived for most of her early life in suburban Detroit, but she imprinted on the wide, sweeping landscapes of Michigan's "thumb" district, several hundred miles to the north, where her family had a large dairy farm on the shores of Lake Huron and where she spent nearly all of her childhood summers. To this day, she feels most at home in places where the north wind roils the surface of the big lake and keens in the windrows of white pine and cedar, where birch logs crackle in the fireplace on chilly August mornings, where the corn fields stretch away to the horizon and the sky is as big and free as it is from the deck of a ship in mid-ocean.

Certain Newfoundlanders I've met feel this same kind of attraction to the geography of their remote outports. For some, whose fathers and grandfathers and great grandfathers may have lived on the shores of the same rocky cove for a dozen generations, the attraction is so strong that no other location in the world—not even one that looks and feels "approximately" similar—can exert the same influence over them. For these villagers, even when forced by a collapsed fishery and a bankrupt economy to board up their houses and move to the far corners of the North American continent, the memories of their traditional outport settings continue to haunt them and their desire to return remains indelibly fixed in their consciousness. In many instances this desire is so strong that even after the last family has moved away and the village has been left to crumble into ruin, its former inhabitants return, year after year, to camp on the broken foundations of the old home sites and participate in week-long "coming home" celebrations.

Sometimes the effect of this kind of geographical imprinting has less to do with the desire to revisit specific settings and more to do with the way such settings influence one's mental imagery and inform the way one thinks about spatial and geographical reality. Noted educator and feminist Jill Ker Conway, in her memoir *The Road from Coorain*, describes her discomfort the first time she left her family's sheep station in western Australia to travel to England, where she was planning to continue her graduate education. She had read about the bucolic glens and gently rolling hills of the English countryside in literature but had never experienced them firsthand. Once there, she quickly realized that for years she had been entertaining the wrong mental picture of these landscapes. Actually seeing them, she admits, was a disappointment. "The light was too misty, the air too filled with water. The Cotswold hills, the deer grazing at the park at Knole, even the great barren that inspired Hardy's Egdon Heath...seemed on the wrong scale."

As a child of the Australian desert, raised and educated in an arid land where the nearest neighbor was two hours away by automobile and the horizon stretched a hundred miles in every direction, she had imagined the geography of rural England on a far more expansive scale. When she actually found herself standing on a grassy hillside in Kent or Sussex or Lancashire, she felt claustrophobic. Everything seemed far too small and closed-in, and she yearned for what she terms "a longer perspective on things."

"It took a visit to England," she explains, "for me to understand how the Australian landscape actually formed the ground of my consciousness, shaped what I saw, and influenced the way a scene was organized in my [mind]. I could teach myself through literature and painting to enjoy this landscape in England, but it would be the schooled response of the connoisseur, not the passionate response one has for the earth where one was born. My landscape was sparer, more brilliant in color, stronger in its contrasts, more majestic in its scale, and bathed in shimmering light."

It would seem to make sense, based on these and other similar illustrations, to suggest that people who have been deeply affected by a specific geography during a formative period of their lives will carry this affect with them, sometimes as an idealized landscape, sometimes as a powerful image of "home," sometimes as an organizational *gestalt*, framing and defining the way they think about the physical world.

But what about the rest of us? What about the people who may not have had a definitive experience of landscape? What about the people (and I count myself among them) who may have grown up in a rather generic urbanized setting that looked and felt like thousands of other generic urbanized settings, strung together with housing developments and cluttered with shopping malls that looked and felt like thousands of other housing developments and shopping malls?

Some who have found themselves in this predicament have decided simply to take matters into their own hands, seeking out the geography that moves them most deeply and reconfiguring their lives so as to live within it. Such was the case with Kathleen Norris, a poet from New York City who found herself yearning for an experience of the vast and barren plains of the western Dakotas, a place she calls "the American desert." It would be inaccurate to suggest that Norris had no previous connection with the landscapes of the great plains, for even though she had not lived there herself, her mother had grown up on a farm near Lemmon, South Dakota, and all four of her grandparents had lived their lives within a few dozen miles of that tiny prairie community.

When the young New Yorker discovered that she had inherited a small house in Lemmon built in 1923 by her mother's parents, she decided, along with her husband, "to make the counter-cultural choice and to live in what the rest of the world considers a barren waste." She chose to leave the vital synergies of New York City and live instead in an isolated and empty land of blistering summer heat waves and howling winter blizzards, a vast ocean of grass so sparsely settled that a large portion of it, with a density of less than three human beings per square mile, fits the commonly accepted definition of wilderness. "Our odd, tortured landscape terrifies many people," writes Norris. "Some think it is as barren as the moon [while] others are possessed by it."

Norris, clearly, is one of those who is possessed by it. In *Dakota: A Spiritual Geography*, she describes her twenty-year encounter with the great western plains as "an experience of the holy." She compares the silence of the prairie to the silence of a monastery, "an unfathomable silence that has the power to reform you," and she describes her own relentless reformation as a kind of spiritual cleansing. The plains, "bountiful in their emptiness," become her teacher. The lesson they teach is a kind of joyous asceticism, a "desert wisdom" that enables her to rejoice in the deprivation of her surroundings and to seek for "that which refreshes from a deeper source."

Odd as it may seem, I too have discovered a spiritual geography that I have come to think of as my own—a rugged and barren sub-Arctic geography that bears little resemblance to any of the places I have actually lived. The city of my birth, a mid-western industrial town called Toledo, Ohio, occupies no space at all in my childhood memory, for I was barely nine days old when my mother packed me in a wicker bassinette and carried me aboard a streamlined passenger train called The Meteor for the ninety-two mile trip to the nearby city of Cleveland. My father, who had been working in Toledo in an experimental metallurgical lab, had just been promoted and asked to move to his company's Cleveland headquarters. Eager for the opportunity to advance in his career, he had agreed to relocate to a setting he did not otherwise know, without relatives or friends or roots of any kind. He and my mother found a small house in an upscale Cleveland suburb called Shaker Heights, and there they set about to create, out of whole cloth, a life for themselves and their young family.

Based upon the eighteen years that I subsequently spent growing up in this setting, it might be reasonable to assume that the landscapes that are imprinted most strongly in my consciousness, the landscapes that feel most like "home," are the flat, monotonous Devonian sea floors and the deeply eroded stream and river beds that together form most of the topography of northeastern Ohio. Yet the truth is that, aside from being familiar, few of the Ohio landscapes have ever felt particularly moving or emotionally satisfying to me. Just as I've tended to feel myself as something of an outsider to Cleveland's human landscape—a visitor, passing through, who may have lived there for a time but was never really *from* there—so I've tended to feel myself as essentially disengaged from its physical landscapes as well. As I've described elsewhere in these essays, I used to love wandering as a boy into the gullies that perforated the edges of our town, for these seemed wild and untamed places, fundamentally different from the trimmed and manicured neighborhood upon the "heights" where my family and I lived. When I finally left Ohio, I carried the memory of the gullies with me, not because they represented some sort of idealized landscape, certainly not because they felt safe or homelike, but because they seemed to point to somewhere else, some as-yet undiscovered place where I might find a spiritual connection with the Earth.

Like Kathleen Norris, I yearned to find a personal geography. But unlike Norris, I had no idea what or where that geography might be. The young New Yorker had the advantage of her grandparents' history, her mother's stories and reminiscences, and, finally, her own (inherited) house on the western Dakota prairie to guide and focus her search. After leaving Ohio, I had six more years of schooling and another ten of secondary school

We were finally able to point our bows in earnest toward the high northern latitudes.

teaching, all in urban or suburban settings in northeastern American cities, where the look and feel of the urban landscapes were little different than in the city of my youth. But I'd also managed to acquire a sailboat—or rather a series of sailboats—and, with my wife Kay and our three boys, I was able to set out each summer to travel under sail, first to familiar destinations along the New England coast, then to more distant and exotic destinations, until the travel began to feel like a quest.

Slowly, inexorably, the direction of these travels shifted northward, until, with our boys grown and off on their own, and with a new and stronger boat, we were finally able to point our bows in earnest toward the high northern latitudes. The attraction, for me, was almost instantaneous. Here at last was a geography I could relate to—a landscape much like the northern ocean itself, that filled you to overflowing with its silences, soothed you with its somber countenance, humbled you with its vast size and scale. It was a landscape almost perfectly suited to a person of my temperament—a landscape that inspired fear and invited contemplation both in the same instant—and I was drawn to it like a moth to a flame.

But this generalized attraction is only half the story, for something else happened during the summer of our first northern voyage that I have never been able to fully understand or explain. One blustery afternoon in early July, as *Brendan's Isle* made her second major landfall of the summer at Sumburgh Head, at the southern extremity of the Shetland Islands, I was overcome with a strange feeling that I had been there before. The bald headland of gray mottled granite, the seabirds nesting in its dark crevasses, the gray-green seas, angry and confused, that swirled beneath the sailboat in almost-whirlpools, the distant shape of a lighthouse atop a lonely promontory, the muted browns and umbers of the land beyond—all these details of what should have been a new and as-yet unseen landscape felt uncannily familiar, as if I had made landfall on this headland many times before.

At first I tried to convince myself that this feeling was some sort of trick my memory was playing on me, a *déjà vu* of some sort, compiled out of a generalized recollection of other rocky headlands I had seen and other remote island landfalls I had experienced over the years. But as we doubled the cape and began making our way along the island's rugged southeast coast, the feeling actually grew stronger. I *had* sailed along this coast before—I was certain of it. The stone croft nestled into the valley, the sheep grazing in the meadow, the circular stone cairn on the hillock to the east, the pinnacle rock jutting out from the shore, each passing landmark re-enforced my feelings of intimacy with this place, as if I were sailing down the last few miles toward home.

I didn't tell anyone on board about these feelings. They were far too strange, and I didn't want to give my shipmates the impression that their skipper might have been losing his grasp on reality after too many weeks at sea. Perhaps if I just concentrated on something else, I thought, the feelings I'd been having might lose some of their immediacy and might come to seem less real. Kay and our son Steve had been traveling by air during the past twenty-four hours from the United States via London to the Shetland Islands. They were due to meet the sailboat in a few more hours on the municipal quay at Lerwick, a meeting we'd been planning for many weeks. I was excited about seeing them after nearly a month at sea. Now, I decided, I would turn my attention to our upcoming rendezvous and try to forget about these odd feelings of mine for a while.

This mental ploy worked for exactly three days. The rendezvous with Kay and Steve evolved into a joyous reunion. The time ashore in the village of Lerwick soon became equally joyous: when we weren't exploring the town and sampling its entertainments, we were fashioning necessary repairs, replenishing food stores, and searching out sources of fresh water, cooking gas, diesel fuel.

On the morning of the fourth day, with the wind rising fresh from the southwest and a wall of fog standing just beyond the harbor mouth, *Brendan's Isle* and her new crew set sail from Lerwick to explore some of the remote northern anchorages of the Shetland archipelago. No sooner had we pulled away from the town quay, however, than the fog rolled in from seaward, forcing us to proceed for the next several hours entirely by radar, without visual aids of any sort. The currents along the route that I'd chosen were strong and unpredictable; the route itself was cluttered with unmarked shoals and ledges. I had sailed in fog along unfamiliar coasts many times before, always with a certain degree of anxiety. Yet, oddly, as we proceeded through a gray, cottony world that morning, I felt singularly calm and unperturbed, as if I'd been sailing blindfolded among these islands all my life.

Shortly after noon, with the sound of a diaphone bleating its sad warning somewhere off to port, we doubled one last headland and turned toward the northern perimeter of a small cluster of islands, identified collectively on our chart as the Out Skerries. Finally, with the crash of breaking waves growing louder by the second, we entered a narrow gash in the rock, then felt our way along a bold shoreline into a large basin surrounded by the spectral shapes of islands on every side.

During the next several hours, with the fog swirling in dense patches through the anchorage, we were afforded a few brief glimpses of the tiny human settlement here. Tucked into the hillside of a small island on the south side of the harbor stood a single wood-frame cottage and a tiny stone church. On a larger saddle-shaped island several hundred yards to the north stood a trio of fishermen's houses, each burrowed into a berm in the hillside to protect it from winter gales. A series of wooden foot bridges connected these larger islands to several smaller ones, each with another house or two nestled into its lee side, while beyond the bridges several groups of open dories lay on their sides in the mud and grass along the shore, like soldiers resting after battle.

Once again, as had happened so often during the past few days, I felt a haunting sense of familiarity with this place—not concrete enough to count as actual memory, perhaps, but deeply compelling nonetheless, as if I might once have visited these islands in a dream. As I stood on the deck of our sailboat gazing about the harbor, I began to entertain a bizarre thought: perhaps in some distant time, ages and ages past, these islands might have actually been my ancestral home, recorded for ever after in the complex twistings of my DNA. Perhaps I *had* been here before, in the guise of my genetic forebears, and this fact had somehow been indelibly imprinted deep in the inner cortex of my brain.

A few days earlier in this journey I would have kept such thoughts to myself. They were, at the very least, impossible to verify empirically, for they posited an interior reality that had no basis in actual experience. This time, however, I decided I'd try to describe them to Kay. I told her about my feelings during our landfall a few days earlier at Sumburgh Head, about the intimacy I'd felt with each passing landmark, all the way into Lerwick harbor. I told her about the confidence I'd felt this morning during our foggy passage to the Out Skerries and about the familiar, dreamlike quality of the Skerries themselves.

"Do you think there could be such a thing as genetic memory?" I asked. "Not just for fish...but for human beings, too? I mean the salmonids have displayed this sort of capacity for millions of years—why couldn't we?"

Kay looked at me with a quizzical expression, and I knew that even as she appreciated the sincerity of my feelings, she was finding my speculations hard to integrate into her unsentimental and firmly practical world-view. "These are beautiful islands," she said at last. "Hauntingly beautiful… I can understand why you love them so."

"But it isn't…I mean it's more than just *love* them," I stammered. "I feel I've *been* here, right *here*, anchored in this harbor. I feel like I might once have actually known the people living in those houses…like I might have walked the beaches on the far side of those hills…."

Kay laughed a gentle, understanding laugh. "Well okay… if you insist. But just promise me one thing."

"Sure, anything."

"Promise that tomorrow you'll introduce me to a few of your oldest friends here…and afterwards that you'll take me to see some of the beautiful places that have meant the most to you over the centuries." She placed her hands lightly on my shoulders and laughed again. This time I found myself laughing along with her.

In the two decades since that first visit to the Shetland Islands, *Brendan's Isle* and her various crews and I have made two more passages across the top of the North Sea, both times pausing to call at that windswept archipelago. The first of these encounters took place less than a year later, in early June, 1985, during a voyage from Bergen, Norway, to the Moray Firth in northwestern Scotland. The second took place near the end of a long sequence of North Atlantic passages along the so-called "stepping-stone" route in the summer of 2000. Both became occasions for lengthy visits to the Shetlands and their environs, and both were marked, for me, by the same prescient feelings of familiarity with the island landscapes.

Kay's gentle corrective notwithstanding, I still found myself entertaining the idea of some kind of genetic imprint at work in all of this. How else was I to explain the attraction that I felt toward this place? The brown sub-Arctic terrain was barren and empty. The meadows were grazed to almost-nothingness by countless generations of sheep. The tallest living plants were miniaturized trees and shrubs, tucked into crevasses in the rock to keep the wind from tearing at their roots. The hillsides were strewn with boulders scattered like skipping stones by the gigantic seas that attacked this place during winter gales.

Much like Kathleen Norris's grassland, the appeal of this landscape resided in its emptiness, in its deafening gales and terrifying silences, and in the utter simplicity of its accoutrements. It was a place unlike any in which

I had ever lived, yet it exerted an irresistible magnetism for me. Each time I returned, the feelings of intimacy grew stronger. Although I never talked about these feelings with Kay (or anyone else) again, I learned to accept them, uncritically, for what they were to me: a signal that I had found the landscape I had been seeking. At last, in this most unlikely of settings, I had discovered my own spiritual geography. I had found the place on Earth where my heart soared and where every sinew in my body told me I was home.

Part III

Time and the Rock

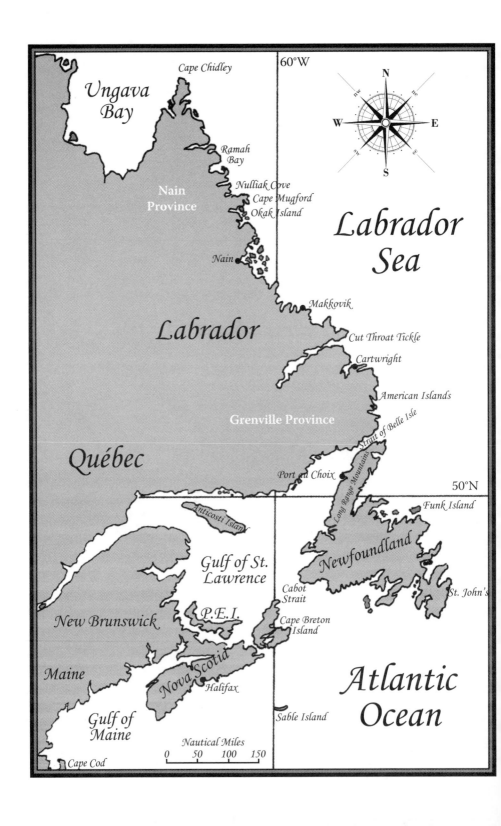

*A brown land dark against the evening sky, treeless and
immensely bleak, and the bared outcropping strata of
the rock stood like a sea wall of coarsed stone; of blocks
so huge that, unconsciously conceiving of mankind by
their scale, we found ourselves immeasurably small.*

Rockwell Kent, *N by E*

12 Messages in the Earth

EAST AND NORTH OF THE Saint Lawrence Gulf on the Atlantic coast
of Labrador, the barren rock predominates. The true tree line and the
domain of the tundra meet the coast in the vicinity of 56 degrees north
latitude (or somewhere near Cape Mugford and the abandoned Moravian
mission station at Okak Island). But the frigid waters of the Labrador Current create their own tree line—one that parallels the coast and transforms
the promontories and near-coastal bays and sounds into a true Arctic eco-
system, all the way down to the Strait of Belle Isle and the great northern
peninsula of Newfoundland.

The experience of sailing anywhere along this coast is an experience of
moving through an exposed and barren terrain. It is impossible to spend a
month or a week or even a day immersed in such a landscape without being
overwhelmed by its scale, humbled by its vast impersonality, mesmerized
by the drama that is written everywhere in its gnarled and cankered visage.

In the ancient terrains of the northeasternmost United States, the story
told in the rock is mostly hidden from us. (Mercifully hidden, one might
suggest.) For here the ancient orogenies have been eroded and smoothed
by wind and rain and the flowing of time. They have been pulverized into
billions of microscopic bits by the roots of two hundred million years, layered over by the ooze of sea beds and the rank fecundity of forests. In these
eroded terrains, we rarely even see the remnants of the rock—unless it is
for the scoured and tumbled boulders scattered across the landscape by the
most recent glacier—or the fringe of naked granite that the sea has scrubbed
clean along the coasts of northern Massachusetts and eastern Maine. But for

What is the composition of these rose-colored headlands?

all the rest it is loam and sand, plowed fields and marsh grasses, hedgerows and wooded thickets, remnant forests on rounded hills and hollows.

As you watch from the deck of your sailboat during a summer voyage to Labrador, however, all of this begins to change. Once having transited the Cabot Strait, you sail for almost three hundred miles along the western coast of Newfoundland, and here you witness a dramatic transformation in the landscape. For the first third of this journey, from Cap Anguille to the palisades of the Port au Port Peninsula, the coastal mountains that form the backbone of western Newfoundland rise precipitously from the sea, covered all about their flanks and foothills by boreal forests of black spruce and pine, hemlock, larch, and tamarack. For the next third of the journey, from the Bay of Islands to Point Rich, the skyline of smooth bald rock that rises above the canopy of green begins to break apart, perforated by huge, eroded gorges that open like massive wounds in the face of the land. The mountains withdraw from the coast and the trees retreat inland and begin to diminish in size. Soon the high coastal terrain gives way to broad outwash plains scattered with boulders and glacial till, covered with coarse beach grasses, and spotted with dwarfed shrubs rooted to the thin soil in gullies and potholes, wherever they can find adequate footholds to anchor themselves against the wind.

Finally, in the last third of the journey as you cross St. John's Bay and enter the Strait of Belle Isle, even these poor remnants of the boreal flora disappear from the landscape, to be replaced by an entirely new regime of Arctic ground cover: mosses and lichens, pitcher plants and trilliums, bogs of broom crawberry, hillsides of bake apples, meadows of miniaturized summer wildflowers. On the southeastern side of the Strait—the New-foundland side—the land falls away to a flat and nearly featureless plain.

But on the northwestern side—the Labrador side—the coastal headlands rise up out of the sea to confront you like the battlements of an ancient stone fortress, each one defining the seaward boundary of another shallow roadstead: Blanc Sablon, L'Anse au Clair, Forteau Bay, L'Anse au Loup, Barge Bay, Red Bay, Chateau Bay, Kennedy Bight.

Once your sailboat has passed through the Strait and onto the southern Labrador, you proceed along a rockbound coast that becomes increasingly more angular and crenellated. Here, confronted at last with a terrain clothed in nothing except its own elemental nakedness, you are forced to ponder the rock. What is the composition of these rose-colored headlands? What forces have formed them and how have they come to lie in their present configuration? Why is the rock that crowns the top of this island layered horizontally while the strata just beneath stand vertically on end? Why are the mountains that form the western skyline crumbling all along their seaward faces, while the ones just to the north rise in glistening black columns that seem impervious to the erosive forces of rain and wind?

Why, why, why? Such were the questions I found myself asking the first time I sailed to this grand and lonely coast. Despite the obvious things I knew (or could surmise) about the ancient history of this place—its location on a tectonic edge where continents had collided and mountain ranges had grown and crumbled—I realized that I was essentially ignorant of even the simplest chronology of geological events that had happened here.

I knew in a vague way about the latest glaciation—the mile-deep blanket of ice that had lain upon this land for twenty-five thousand years—for this was the same ice that had covered northeastern Ohio, northern Pennsylvania, most of New York state, and all of New England down to the southernmost terminus of the ice along the moraines of Long Island, Martha's Vineyard, Nantucket, and the drowned islands of the Georges Bank. I had learned about this ice as a boy, for the Ohio landscape where I had spent my childhood was a treasure trove of glacial detritus, ice-carved gorges, and the ancient graveyards of shaggy ice-age monsters.

Beyond these boyhood impressions of a frozen world that had ended some ten thousand years ago (in what I then perceived as the ancient mists of time), I had almost no mental picture of the hundreds of millions of years that had preceded this, our modern era. In the absence of the naked rock to challenge my ignorance, I was content to live my daily life as if the Earth I knew had never changed, as if the continents had always occupied

An unusual formation known as the "Bishop's Mitre," near Cape Mugford, northern Labrador.

the places on the globe where they presently resided, as if the oceans had never opened into ever-widening cataracts or flooded onto continental plains or disappeared altogether as the seafloor underneath them plunged into subduction zones and cycled back into the planet's molten center.

As soon as I returned from that first voyage to southern Labrador, I promised myself that I would end my ignorance, to the extent that I was able, and learn the geologic history of this place. I had an old friend from my high school days back in northeastern Ohio, Steve Stanley, who had focused much of his professional career around the study of the Earth's history as it is written in the rock. Over the past few decades, Steve had moved from Princeton, where he had completed his undergraduate studies, to Yale, where he had earned his doctoral degree in paleobiology, to Johns Hopkins University in Baltimore, where he had served as chairman of the Department of Earth and Planetary Sciences, and finally to his present position as Professor of Geobiology at the University of Hawaii.

I glanced through a handful of Steve's books and scientific articles—just to apprise myself of his current interests and learn about some of the subjects of his research—then I made a telephone call to his home one Saturday morning to ask him for his help.

"How would an old sailor and a once-upon-a-time high school English teacher like me learn about the geology of a place that almost nobody cares about and fewer have ever visited?" I asked.

Steve laughed. "Well it seems like you're in luck, my friend. That coast you're referring to happens to be smack in the middle of one of the most active geologic provinces on the North American craton. There's probably been as much tectonic activity going on in that particular region during the past billion or so years as there has been anywhere else on the planet: the creation and dismantling of two supercontinents, the forming and flooding of two major oceans, three mountain-building episodes, the repeated opening and filling of a foredeep, the suturing and breaking apart of half a dozen continental plates. The chronology of how it all happened ought to make a pretty good story."

Steve promised to send me a copy of a book that he'd written, *Earth and Life through Time*, and that he had used at Hopkins to introduce his students to the study of geology. He advised me to study the text—just as if I were one of his students—then to contact him again if I needed more help. The story that follows is the result of this study—and of several additional sessions with a man who is recognized by his peers as one of the most brilliant and creative minds in his field. The story, Steve has asked me to emphasize, is pieced together like a giant jigsaw puzzle from fossil evidence, radioactive dating, and structural analysis of rock and sediment formations, as gathered by researchers working all over the planet. It is a story consistent with the evidence as we know it so far—but, just as with all good science, it is necessarily also a work in progress, subject to change as our understanding of Earth systems becomes more sophisticated and the data sets become more complete.

The story begins during the Proterozoic era, approximately two billion years before the present (BP), when a number of the lithospheric plates that carry the continental bedrock on its journey about the surface of the planet started drifting closer together. Slowly, over the next few hundred million years, these plates began colliding with one another, suturing plate on plate, until they formed a single, massive, continental land mass: the so-called Proterozoic Supercontinent.

By the end of this continent-building period, about one billion years BP, the bedrock that underlies much of present-day Labrador achieved a configuration roughly similar to the one it has today. Even then this basement rock was comprised of two distinct geologic formations: first, the so-called Nain Province, a formation that had become amalgamated into the North American craton close to two billion years BP and that included the northern and western portion of present-day Labrador; and second, the so-called Grenville Province, an area of exotic terrain added by accretion all along the eastern edge of North America slightly more than a billion years ago and that included present-day Gaspe, southern Quebec, and southern Labrador, along with much of the great northern Peninsula of Newfoundland. By about one billion years BP, both the Nain and the Grenville Provinces lay deep within the central region of the Proterozoic Supercontinent, with

North America to the west, Greenland, Siberia, and northwestern Europe to the north and east, and Gonwandaland (either South America or North Africa—the geological evidence is unclear) to the south. Far from being a coastal terrain, these two geologic provinces of the "proto-Labrador" were located as far away from the sea as, say, Minnesota is today.

About 800 million years ago, during the late Proterozoic era, a rift began to re-form along the old boundary line between North America and Gonwandaland. Soon the rift area became flooded by seawater—the beginning of the Iapetus Ocean—leaving the rock of the Grenville Province as the eastern perimeter of the new continent (and rendering the basement rock of southern Labrador into waterfront property once again).

As the Proterozoic Supercontinent continued to break apart over the next 200 million years, Gonwandaland moved farther and farther off to the south. Greenland remained sutured to North America in the vicinity of northern Labrador, as did Scotland and Northern Ireland, forming the newly amalgamated continent of Laurentia. Siberia and northern Europe rifted apart and began drifting off to the north and east.

By the beginning of the Ordovician period (about 570 million years BP), the Nain Province was still emplaced deep within Laurentia, forming the eastern end of the so-called Canadian Shield (a massive area of ancient felsic crust that together comprised the basement rock of the North American craton). Most of the Grenville Province, meanwhile, had become flooded by a shallow sea all along the eastern rim of the newly assembled continent. This area included all of southern Labrador as well as most of the great northern peninsula of Newfoundland. To the west and south of this area, a deep ocean trench or "foredeep" formed in what would one day become the Gulf of Saint Lawrence and western Newfoundland. Deep-sea deposits of muds and brec-cias slowly filled this feature during the next 100 million years, as Laurentia drifted slowly southward and the Iapetus Ocean continued to widen.

Finally, about 475 million years ago, during the mid-Ordovician pe-riod, Laurentia had wandered so far south as to become a tropical conti-nent. The equator passed directly through present-day Cape Chidley (at the northern tip of Labrador) and Cape Farwell (at the southernmost tip of Greenland). The basement rock of present-day Newfoundland was part of a shallow, tropical sea lying in about 10 to 15 degrees south latitude (or about the position of northernmost Australia today.) And just offshore, an arc of active volcanic islands slowly approached from the east as the ocean floor was subducted beneath them—harbingers of massive changes about to take place all across this region.

Between the mid-Ordovician period (470 million years BP) and the late Carboniferous (286 million years BP), three distinct episodes of mountain-building erupted along the rift and fracture zones of eastern and northern Laurentia. The first episode—the so-called Taconic orogeny—burst upon the landscape during the middle Ordovician period as a result of the near-approach of the northern European craton of Baltica and a subsequent collision between Laurentia and an island arc that lay entrapped between the two. During this collision, huge blocks of deep-sea floor were emplaced onto the original (Grenville) craton, forming a mountainous uplift that is still visible in the highlands of western Newfoundland: the Bay of Islands, Gros Mourne, the Long Range Mountains, and the Highlands of St John. To the south and west of these features came the rest of the Taconic orogeny, a mountain-building episode that resulted in a nearly continuous range of mountains along the eastern margin of Laurentia. Adjacent to the uplift in the vicinity of western Newfoundland, another foredeep was formed; to the east a band of basaltic seafloor was accreted from the island arc and emplaced on the original bedrock to become the middle section of present-day Newfoundland.

After an interval of relative quiet during the Silurian period (438–408 million years BP) during which the Taconic mountains were eroded to near-oblivion, a second major mountain-building episode began to take shape. The two-pronged catalyst for this event was the collision and suturing of Baltica and Laurentia in the north (forming what geologists term the Old Red Sandstone—or ORS—Continent), and the collision of Laurentia with yet another island arc in the south. The result was a mountain-building event of massive proportion, the so-called Acadian (or Acadian/Caledonian) orogeny.

As the ORS Continent drew together, the Iapetus Ocean withered in size and finally closed altogether. England and southern Ireland became sutured to Scotland and Northern Ireland in the center of the new continent. The Acadian uplift, in the form of the proto-Appalachian mountains of eastern North America, was joined at its northeastern end by the Caledonian mountains of Scotland, the mountains of eastern Greenland, and the mountains of western Norway, making it one of the longest continuous regions of mountainous uplift on the planet since Archean times. In the interior of this continent, high on the western slope of the mountains and very near to the earth's equator, lay the tropical highlands of southern and eastern Labrador. To their south, another large area of crustal bedrock, the so-called Avalon Terrain, had rafted up onto the Grenville, forming what would one day become eastern New England, the highlands of Cape Breton Island, and the Avalon Peninsula of southeast Newfoundland.

The third major episode of mountain-building affecting the proto-Labrador began about 300 million years ago during the late Carboniferous period. Like the first two, this episode also took place as a result of the collision and eventual suturing of large areas of crustal bedrock—in this case, the ORS Continent with Gonwandaland. This merger became the main event in the creation of Pangaea, an amalgamation of virtually all the dry land on the planet and the second great supercontinent. Pangaea was completed with the suturing of Baltica and Asia/Siberia near the end of the Carboniferous period (about 285 million years BP), after which time the entire supercontinent became surrounded by a single immense body of water: the so-called Panthalassic Ocean.

As Pangaea was being formed, a new generation of mountain ranges arose along the boundaries where the crustal plates collided. These included the Alleghenian orogeny, a mountain-building episode that took place along the same tectonic edge that had hosted the Taconic and Acadian orogenies several hundred million years earlier. The result, in the vicinity of the proto-Labrador, was yet another area of mountainous uplift deep in the interior of the new supercontinent. During this third orogenic event, another foredeep (the proto-Gulf of Saint Lawrence) was formed in the vicinity of western Newfoundland, where it persisted for tens of millions of years in the guise of a large inland sea.

The breakup of Pangaea, beginning about 240 million years BP, served in many ways as the signal event announcing the onset the modern geological era. As the African and North American continents began drifting apart again in the early Triassic period, the waters of the so-called Tethyan Overflow began spilling westward in sporadic bursts along a fracture zone that was opening between Europe and Africa (and that would eventually become the Mediterranean Sea). By the late Triassic (200 million years BP), the fracture zone had grown to include the eastern coast of North America, and by the early Cretaceous (130 million years BP), sea water from the Tethyan Overflow had begun flooding this entire region, creating a long, narrow basin of sea water: the proto-Atlantic Ocean.

The radical folding of the "Arctic cordillera" is everywhere apparent in Labrador's Tourngat Mountains.

By this time, Newfoundland had almost certainly become an island again, with the faulting and rifting on the eastern (Atlantic Ocean) side combining with the flooding of the ancient inland sea on the western (Gulf of Saint Lawrence) side. With this flooding, southern Labrador was also transformed into a coastal landscape once again.

Northern Labrador, meanwhile, was still sutured to Greenland and northern Europe, awaiting the final breakup of Pangaea before it would enter a period of dramatic folding and faulting (the so-called Arctic Cordillera), culminating in yet another orogenic event—the Innuitian orogeny. At the completion of this final mountain-building episode, and with the continued northward flooding of the proto-Atlantic basin, northern Labrador was transformed into a coastal range of high peaks and steep-sided fiords—a terrain still visible today as one sails north from the vicinity of Okak and Cape Mugford into the present-day Torngat Mountains.

By the late Cretaceous period, about 70 million years BP, all the major continents had achieved positions on the surface of the planet roughly similar to the ones they occupy today. The Atlantic Ocean was still growing wider as new sea floor continued to erupt from volcanic vents all along the mid-Atlantic ridge. Africa had rifted away from South America and had taken up its current position on the eastern side of the Atlantic. South America had drifted westward and had become connected by a narrow land bridge with its sister continent to the north. Greenland had broken away from Labrador, creating the Labrador Sea and Davis Strait and opening up the seaway that would eventually become the Arctic Ocean.

By the beginning of the Eocene epoch (about 58 million years BP), the waters of the North Atlantic had become an ideal habitat for the evolution of the earliest whales. By the early Miocene (about 24 million years BP), the first apes and monkeys appeared in the rain forests of equatorial Africa. And finally, by the beginning of the Pleistocene (about 1.65 million years

BP), one of the earliest precursors of the human family, *Homo habilis*, arrived upon the scene, completing its evolution from its ape-like ancestors in response to a series of major glaciations that had started advancing all across the northern hemisphere.

Another 1.45 million years would be required, however, before our own species, *Homo sapiens*, would appear upon the evolutionary scene. And tens of thousands of years more would pass—all the way down to about 9000 years BP—before the earliest waves of human hunter-gatherers would arrive at the northeastern edge of the American continent, gaze out upon the ice-polished mountains of southern Labrador and western Newfoundland, and set foot at last upon this frigid, rockbound shore.

Thus ends the story written in the rock—the story I first learned from my friend and mentor, Steve Stanley, and the one I've been re-telling ever since to shipmates who have accompanied me on voyages to this place.

The story accomplishes two important purposes, I think, for those who choose to pay attention. First, it provides a context within which to pose our questions. It gives us a framework—a kind of "table of contents" for the encyclopedia of the Earth—instructing us about the layering of events and the profound depth of geologic time.

Second, the story teaches us an essential lesson about ourselves, for it stands as a blatant challenge to our convenient formulas of life and death, time and immortality. It is a story that has taken more than two billion years in the telling, yet our portion of it is barely two hundred thousand of those years—a time so short as to be virtually invisible in the geological record. What this means is that we, in our entire history as a species, are not yet even a "sudden event." We have not yet claimed a single layer in the lexicon of the Earth's genealogy. To become successful, to merit a fossil record worthy of any future attentiveness, we will need to flourish and prosper upon this planet not a thousand years more, not a hundred thousand, not a million, but twenty times a million years. To achieve the evolutionary status of, say, the long-finned North Atlantic pilot whale, we will need to prove our staying power for twice twenty million years.

Thus chastened, my shipmates and I turn our attention to the rock once more as we slowly make our way along this coast. Gazing up at the architecture of a thousand million years, we wonder anew at its vast chronologies and nearly incomprehensible time scales, and we shudder at the intimations of our own inconsequence.

*Environmental damage, climate change, rapid
population growth, unstable trade partners, and
pressure from enemies were all factors in the demise
of doomed societies [even as] other societies found
solutions to those same problems and persisted.*

Jared Diamond, *Collapse*

13 Ancestors

ON THE OLD FRENCH SHORE of Newfoundland stands a dramatic
promontory, the Point Rich Peninsula, that juts out three miles into
the northern Gulf of Saint Lawrence. To the west of the peninsula is a deep
channel—part of the remains of the old Saint Lawrence foredeep—where
strong currents upwell and the skate, dogfish, pollack, and ocean perch
feed in dense aggregations. On the north side is a series of terraced stone
beaches where herds of harp seals pause to feed and suckle their young
during their northward migration each spring. And on the east side, tucked
under the lee of grassy hills, is the tiny fishing village of Port au Choix, built
atop an ancient raised beach and facing one of the few protected natural
harbors in northeastern Newfoundland.

The modern history of the Point Rich Peninsula began at the turn of
the twentieth century—1904, to be precise—when a new fishing treaty between France and England opened the territory to settlement by English-
speaking colonists. During the next several decades, the village of Port au
Choix became a frontier boom-town. Fishermen from all around Newfoundland began migrating north to try their luck in the newly opened
fishing grounds, and the ones who decided to stay were soon building new
homes and storehouses all along the village's main street.

Each time ground was broken for a new foundation, however, a problem would arise. In building site after building site, no sooner did the shovel penetrate the peaty soil more than a few feet than the ground would
appear saturated with a bloody red dye. Another foot or two down the
shovel head would strike something hard—a stone implement, perhaps, or

a highly polished spear point, or the toggling head of a bone whaling harpoon. Next to these (if the shovel didn't crack right through them) would appear a splay of ribs, a shoulder blade, a collarbone, a line of vertebrae—and finally, leering up out of the reddened soil with all its teeth intact, a human skull.

The experience was unsettling, even for the most hardened of the newcomers, for nobody wanted to disturb another person's grave. It was soon determined, however, that the gravesites being opened were not French or English—nor in fact were they any other obvious European type. They must have been Eskimos, the homebuilders decided, buried here long before any European had chanced upon this shore. It was a shame to disturb their graves, to be sure, but what else could be done? Progress was progress, after all, and the village needed to expand....

Nobody in Port au Choix thought much more about the old Eskimo graveyard under their town until one June day in 1967 when a local businessman decided to break ground for a new movie theater in a large open lot near the center of the village. Once again as the excavation began, the soil in a number of the opened areas became saturated with a crimson dye, as if the ground below were oozing blood. This time, as reports of the incident began to circulate, a young archaeologist named James Tuck, from Newfoundland's Memorial University, happened to hear of them. Tuck had been interested for a number of years in the ritual practice of red ochre (or "red paint") burial among ancient aboriginal civilizations, and he was eager to travel to Port au Choix to get a closer look at these potential red paint burial sites before the ground around them could be further disturbed.

During the following summer, Tuck and a team of assistants excavated three burial locations in the town center, uncovering nearly 100 almost perfectly preserved human skeletons as well as numerous tools, weapons, fishing and hunting implements, ritual artifacts, and other grave goods. Their first surprise was the extreme age of the cemetery contents, for after completing a series of radiocarbon dates, they were able to fix the age of the burial remains at between 4400 and 3300 years BP (before present), several thousand years older than any other then-known human remains in northeastern Canada. Their second surprise was their difficulty in identifying the cultural typography of the cemetery people, for not only were the Port au Choix graves *not* of Eskimo origin but they were also not of any other cultural or ethnographic origin yet known in Newfoundland.

Tuck was quick to recognize the high quality and skilled workmanship of the artifacts found in the Port au Choix graves. He understood that although racially related to the Amerindian, the Port au Choix people enjoyed a culture far superior to more modern Indian cultures. Their polished stone fishing pendants, their beautifully formed woodworking tools, their decorative lances and spearheads, their barbed fishing hooks and toggling harpoons all pointed to a highly evolved culture of skilled hunters and fishermen, well adapted to maritime settings. The red ochre dye that covered the contents of their graves suggested a connection with a series of "red paint" burial sites discovered during earlier decades along the coast of Maine and also radiocarbon-dated to approximately 4000 years BP. Both the Maine sites and the Port au Choix site contained spearheads and other artifacts formed of a material called Ramah chert, a creamy smooth, flint-like substance that was only quarried in one small fiord in northern Labrador. The appearance of artifacts of Ramah chert in both locations indicated that there must have been some form of contact between the two—if not direct, then by means of a sophisticated long-distance trading network—as well as between both southern locations and the miners in the chert quarries many hundreds of miles to the north.

What all this meant was that Tuck and his colleagues had uncovered a new culture, a "lost nation" of maritime-adapted hunter-gatherers never before encountered in the archeological record of the far north. Grouping them with the red paint cultures of the Maine coast, Tuck coined a new name—the Maritime Archaic Indians—thereby emphasizing both the extreme antiquity and the strong maritime orientation of their way of life. Based on the clear signs of successful adaptation of their tools and other artifacts to the living conditions in northern coastal settings, he postulated the discovery of more burial sites—and possibly habitation sites as well—in both northern Newfoundland and east-coastal Labrador, perhaps as far north as the chert quarries in Ramah Bay. In the decades that followed, he, along with a small but dedicated group of other North American archaeologists and cultural anthropologists, set about to find them.

The first time *Brendan's Isle* and her crew and I sailed into Port au Choix was in the summer of 1988. We were on our way to Labrador that year, making our way north along a coast that was still new to us. One morning, with a forecast low-pressure system approaching and the prospect of sev-

Brendan's Isle *safely moored at the government pier in Port au Choix, northern Newfoundland*

eral days of stormy weather in the Strait of Belle Isle, I began searching my charts for a sheltered anchorage where we could stop moving for a day or two and wait for the return of settled conditions. That afternoon my shipmates and I found ourselves watching with anticipation as the silhouette of the Point Rich Peninsula loomed up over the horizon and slowly grew into the form of solid land. Like the thousands of mariners who had traveled these waters before us, we had settled upon Point Rich and the harbor at Port au Choix as the safest and most accessible stopping place for many miles in either direction.

Once the sailboat was safely moored at the government pier, my shipmates and I were free to wander about the village. The first hint I received that there might be something unusual happening here came during a conversation with a shopkeeper down near the center of town. "Have ye seen the skeleton?" queried the man. When I indicated with a shake of my head that I had not, he pointed excitedly toward a cluster of small octagonal buildings on the far side of the road. "Very old it is," he said, "—old as the Kings of Egypt. A young doctor-fella name of Tuck come out here from St. John's some years back and found near a hundred of 'em. Some almost perfectly preserved. You'll see one of 'em lying on a table, almost like he was asleep, over in that little wooden building."

With my curiosity now thoroughly piqued, I stepped across the road to a large open field criss-crossed with a series of wooden boardwalks. Signs bearing the "Parks Canada" logo had been placed at various stations around the field to explain the history of the archeological discoveries here and to summarize what was currently known about the Maritime Archaic culture. Following the boardwalk past a series of grave sites, I came at last to the door of the main museum building and stepped inside.

In the semi-darkness I could just make out the display case that stood near the center of the room. In the case was a human skeleton, reassembled so as to appear in its original burial posture with various grave goods placed about its head, arms and torso. Except for the dark reddish brown color of the bones (a color, I later learned, caused by the red ochre burial paint), the figure looked like the remains of a modern cadaver, recently exhumed. According to the accompanying interpretive material, he was a young male— possibly a fisherman—who had been found with the polished stone effigy of a killer whale placed within the cavity of his chest. Others from the same grave site had been accoutered with bone combs intricately carved in the forms of seabirds in flight, as well as with other decorative ritual artifacts.

During the next several days, as my shipmates and I waited for the easterly gales to subside, I found myself becoming more and more intrigued with the Port au Choix cemeteries and with the race of ancient people who had been discovered here. Who exactly were they? Where had they come from? How had they developed such sophisticated toolmaking technologies and such advanced maritime skills, especially in light of later Amerindian cultures that showed none of these advances?

I discovered a partial answer in a set of interpretive materials displayed in the little museum building at the Port au Choix cemetery site. It seems that James Tuck, after finishing his work here in 1978, had turned his attention to another site that had long interested him in L'Anse Amour, a small bay about sixty miles northeast of Port au Choix on the Labrador side of the Strait of Belle Isle. Here, in the summer of 1980, he had excavated a mound-shaped structure that turned out to be another Maritime Archaic burial site, this one containing both artifacts and skeletal remains carbon-dated all the way back to about 7500 years BP.

This discovery, along with another in nearby Pinware Bay, Labrador (carbon-dated to approximately 9000 years BP), moved the clock for Maritime Archaic habitation in northeastern Canada backwards to a period very near the time when the Wisconsonian glacier was just finishing its retreat across the Strait of Belle Isle. What this meant was that unless these people had actually been able to migrate from east to west across a frigid and barren ice cap (a feat considered by most scientific observers to be all but impossible), they must have come from the other direction, moving from west to east across continental North America, and representing (as Tuck and others have concluded), "the final expansion of Paleo-Indian migration across the New World."

The question of how the members of this ancient culture managed to develop their maritime skills during their long trans-continental migration remains something of a mystery—and is the subject of ongoing scientific debate. One possible scenario suggests that a final sub-population actually reached the Atlantic coast during the late Wisconsonian era somewhere south of the glacial advance, then followed the ice-edge as it retreated, honing and perfecting their fishing and seafaring skills as they slowly moved northward.

Even more compelling than the questions of their origins, I soon learned, were the questions surrounding their demise. For indeed, at just about the time of the last graves in Port au Choix, and after more than 5000 years of successful adaptation to northern maritime environments, the entire southern branch of the Maritime Archaic and their highly advanced culture suddenly disappeared from the face of the earth.

At the time of our first visit to Port au Choix in 1988, nothing had been found in the archeological record that post-dated the final burials. This somewhat perplexing situation persisted even after a new team of Canadian archeologists, working during the mid-1990s, finally located the long sought-after habitation site of the Port au Choix Maritime Archaic community on another raised beach just to the south of the Point Rich Peninsula. Indications of a thriving way of life continued throughout the various layers of this habitation site—just as they continued in both habitation and burial sites discovered around the entire perimeter of Newfoundland—until about 3200 BP. And then, without a clue as to what catastrophic event (or events) might have taken place, the record simply ended.

By the time I became fully apprised of this archeological mystery, several more years had passed and I was busy with another project—a voyage to Greenland to investigate the relationship between ice and climate, especially as these related to current changes taking place within the Arctic sea ice regime and their potential impacts on climate patterns worldwide. As part of my preparation for this project, I was interviewing researchers from a wide range of scientific disciplines—climatologists, oceanographers, paleobiologists, atmospheric chemists, glaciologists, and others—trying to piece together an emerging story that might have relevance to human communities all over the Earth.

One such researcher, introduced by a mutual friend, was an American archeologist then working for the Smithsonian Institution by the name of William Fitzhugh. At the time, Fitzhugh was pursuing summer fieldwork in the vicinity of southern Baffin Island, but all during the previous decade his work had been focused in northern Labrador, where he had been involved in the search for the northernmost habitation site of the so-called "northern branch" of the Maritime Archaic.

A telephone call to Fitzhugh in January 1994 resulted in an invitation to travel down to Washington D.C. for a meeting at his winter headquarters deep in the inner sanctum of the Smithsonian Museum of Natural History. On the day of our appointment, my soon-to-be shipmate Mikey Auth and I made our way to an obscure set of elevators behind one of the dinosaur exhibits, then up several floors to a darkened corridor lined with boxes and specimen cases. Part way down this corridor we found a glass doorway marked with Fitzhugh's name. Inside the door was a large, brightly lit room adorned with maps and diagrams of archeological sites and cluttered with all manner of other archeological oddities. Seated behind a large oak desk was Fitzhugh himself, a strikingly handsome fellow, fit and lean, with a shock of wavy brown hair and the leathery-tan complexion of one who has spent much of his life outdoors.

After attending to the necessary introductions, I turned to the purpose of my visit. I was interested in Fitzhugh's work, I explained, primarily as it related to paleo-climate, for I was looking for possible analogies between the changes taking place in present-day climate patterns and those that may have taken place in the distant past.

"Climate indicators in the far north are difficult to tease out of the archeological record," Fitzhugh remarked, "unless the changes are really dramatic—dramatic enough to alter the geographical range of major flora, for example, or to cause animal or human populations to migrate into new areas." He paused, then began rustling through a file of papers on his desk. "There *is* one climate event that I can tell you about, though, that may be exactly the sort of thing you're looking for...."

With this preamble, he set off on a description of his decade-long investigations of Maritime Archaic presence along the Labrador coast. Beginning in 1974, he explained, while James Tuck and others were continuing their exploration of sites in Newfoundland and the Strait of Belle Isle, he and his associates from the Smithsonian began investigating a series of more northerly sites. As part of the infrastructure for these expeditions, the Smithsonian purchased an old Grand Banks fishing trawler, later renamed the *Pitsiulak*, and had her refurbished in the manner of an expeditionary transport vessel expressly for Fitzhugh's professional use. Early each summer, the archeologist and his entourage of students and support personnel would board this vessel at her winter quarters in eastern Newfoundland. From here they would steam north, following the receding pack ice to a series of increasingly remote exploration sites: Hound Pond, Sandy Cove, Black Island, Rattlers Bight, Cutthroat Tickle, Ballybrack Cove, Gull Arm, Okak Island.

Summer after summer for the next six years, these expeditions pushed ever farther north, following a trail of ancient clues from habitation site to habitation site, burial mound to burial mound, uncovering the story of a new and previously undocumented branch of Maritime Archaic presence. Finally, in the summer of 1980, Fitzhugh and the *Pitsiulak* arrived at Nulliak Cove, a complex of 26 stone foundations in the heart of the mountainous Tourngat region that would eventually prove to be the northernmost known habitation site of this ancient civilization.

Radiocarbon dates fixed the heyday of this settlement at about 4000 years BP, Fitzhugh explained, with indications that it might have been inhabited several hundred years earlier. "It's a fascinating site, with engraved ritual pendants, a mysterious standing stone, all kinds of neat stuff. For your purposes, though, its most important feature is probably its ending—for right about 3700 years BP, in the midst of a cooling event that seems to have dramatically altered the climate in northern Labrador for several hundred years (the so-called 'post-Hypisthermal cold period'), this settlement was abandoned. As the climate deteriorated and living and hunting conditions became more difficult, a paleo-Eskimo group called the Pre-Dorset began encroaching from the north, displacing Maritime Archaic populations as they moved down the coast.

A retreating ice edge of the sort that the Maritime Archaic Indians may have followed as they migrated northward

"All we can assume is that this new Eskimo population was better adapted to the cold conditions and was able to out-compete the older Indian populations for available resources. We'll never know for sure whether there was actual armed conflict or whether the older inhabitants simply became overwhelmed during a sequence of particularly harsh winters. All we know is that within a few years, a civilization that had flourished along this coast for millennia had disappeared."

"This sounds curiously like the story told by Tuck and his colleagues," I said "—about the disappearance of the southern branch communities down along the Newfoundland coast and the Strait of Belle Isle."

"The two events are similar," said Fitzhugh, "—although they're hardly identical. For one thing, the southern branch people seem to have survived the post-Hypisthermal cold period quite successfully. Their communities were still thriving at 3700 years BP when the northern branch communities were falling into their final decline. The southern branch people flourished another 500 years, in fact, all the way down to about 3200 years BP, before their record ended.

"A second difference has to do with the encroachment of the Pre-Dorset people in northern Labrador and the competition for resources that almost certainly followed. Down in Newfoundland, the southern branch people seem to have faced no such competition. The pre-Dorset migrations never extended farther south than the Nain Archipelago. And the next wave of Eskimo encroachment, the Dorset, didn't arrive in Newfoundland until a thousand years *after* the southern branch people had already disappeared."

My shipmate Mikey spoke up next. "Tuck and a few of his associates have suggested that the southern people may not have entirely disap-peared—that a remnant population might have survived to become the genetic forbears of modern Indians in Newfoundland."

"Quite frankly," said Fitzhugh, " I've never found the evidence in the archeological record to support this kind of speculation. Is it possible that a few individuals managed to survive the general collapse? Yes, that much is certainly possible. But we're still left with all the same questions about what happened to the general population and to the civilization as a whole.

"Maybe in coming years a new generation of researchers will un-cover a southern branch site containing the missing clues—the way our team was able to do in the north. Meanwhile the best we can do, as sci-entists and as interested onlookers, is to remain patient in the presence of mystery."

For reasons of narrative economy, these conversations with Bill Fitz-hugh were never included in the book I wrote about ice and climate change. They remained fixed in my memory, however, until I finally found time ten years later to return to maritime Canada and revisit several of the archeological sites where so much of this ancient tale had taken place.

The year was 2004; the time was early July, only a week or so after the last of the pack ice had disappeared from the Strait of Belle Isle. I was trav-eling fast that summer with a new group of sail-trainees, planning to sail as far north along the Labrador shore as time and weather would permit. But before tackling the Strait, I wanted to visit the Port au Choix cemetery once more, see the excavations at the Maritime Archaic habitation site, walk out to the ruins of a Dorset Eskimo whaling station on the north shore of the peninsula.

The first surprise came at the old cemetery site, for search as I might, I was unable to find a single sign of the archeological activity that had marked the place so dramatically sixteen years before. All the graves had been filled in; all the paths and boardwalks had been demolished. The cluster of octagonal museum buildings, although still standing, had been given to a local historical society. The skeleton that had inhabited the glass display case had been removed and re-buried, I was told, at the insistence of several present-day Indian organizations that had decided to take Tuck's speculations literally and to claim the cemetery people as their own.

A brown and yellow "Parks Canada" sign on the main highway just south of town indicated that a new interpretation center had been built several miles out on the Point Rich Peninsula. When I asked a group of local villagers about the new center, I was told that most of what were now on display were scale models and drawings and photographs. Likewise out at the newly discovered habitation site, they explained, all the signs of the actual archeological activity had been removed. The excavations had been filled in and sodded-over—just like the grave sites back in the village—to return them as nearly as possible to their original, undisturbed state.

With the hour growing late, I decided to forego the walk to the new interpretation center or the long trek out to the habitation site. Instead, I set off on a poorly marked footpath toward yet another ancient site, this one an early Dorset Eskimo fishing, sealing, and whaling encampment dating from about 2000 years BP and discovered some years ago by American archeologist Elmer Harp. The path, winding over grassy hills, finally deposited me on a stone-terraced beach known locally as Philips Garden. A metal sign set in cement at the edge of the uppermost terrace announced that I had indeed found the place I was looking for. Beyond the sign, however, there was nothing—no standing stones, no cairns, no foundations, no piles of broken rock where the walls of buildings once had been. As far as the eye could see were only windswept hills and row upon row of empty shale steps descending haphazardly toward a ragged line of surf.

I'm not sure what I'd been expecting—but as I stood there, I slowly came to realize that this blasted and empty landscape was precisely the sort of place in which this story should find its ending. The Port au Choix cemetery people had been returned to their graves. Their elusive and long-buried habitation site was once again covered with soil and grass. And now, I realized, the Dorset hunters and fishermen who had thrived for centuries upon this beach were gone as well, disappeared without a trace.

The Earth had reclaimed them all, and we were left to try to remember their stories, to glean whatever lessons we could from them... or else—who knows?—perhaps to repeat them some day as our own.

Such high densities of suspended calcium carbonate coccoliths can form spectacularly bright, milky, or turquoise-colored patches, spreading from horizon to horizon over hundreds of thousands of square kilometers. These blooms, occurring over geological time, have been responsible for the formation of the white cliffs of Dover...

William M. Balch, *Our Changing Planet*

14 Milk Sea

DURING ALL THE YEARS THAT my shipmates and I have sailed *Brendan's Isle* into the northern ocean, we have witnessed many wonders. We've celebrated the dalliance of pilot whales, marveled at the mirages of upside-down icebergs floating at the edges of a spectral sea, sailed under the bluegreen fingers of the Aurora Borealis as they pulsed across the northern sky. We've seen the sudden explosion of geysers of scalding hot steam erupting from the walls of Iceland's westfjords; we've watched the calving of half-mile-long icebergs from one of Greenland's moving rivers of ice. We've been awed by the searing white glow of a mountain glacier a hundred miles beyond the horizon; we've lived for weeks without darkness in the land of the midnight sun.

Perhaps nowhere during the course of all these encounters, however, have my crew and I been more amazed and confounded than during *Brendan's* first foray into the northern Atlantic in the early summer of 1984. Here, on the morning of June 24, just as the sailboat was reaching 58 degrees north latitude, some four hundred miles south of Reykjavik, Iceland, an unusual entry appeared in the ship's logbook:

Southwest gales, breaking seas, low stratus cloud racing overhead, all familiar now after eight days of running down the North Atlantic storm track. But something odd has happened to the sea this morning. During the past several hours the familiar gray-green color of the ocean surface has been transformed. Now, everywhere we look,

We have witnessed many wonders.

> *the water has taken on a bright turquoise color, as if we were sailing over a shallow bank of sand.*

Nobody on the sailboat knew what to make of the strange circumstance in which we suddenly found ourselves. The brilliant turquoise color of the water seemed to glow with an interior light. The breaking white crests spilled down the faces of following seas like whipped cream. The surface of the ocean was opaque, nearly impervious to the filtered sunlight, so that the water color actually reflected back into the layer of scuddy cloud above. The effect was oddly theatrical, as if a tinted screen had been dropped across a set of stage lights, casting the entire scene into an eerie monochrome of robin's egg blue.

Since neither I nor any of my shipmates had ever sailed before in this part of the world ocean, no one had a base line against which to evaluate the bizarre coloration we were witnessing. As far as any of us knew, this creamy turquoise hue was in fact the "normal" color of the sea water in the cold high latitudes of the northeastern Atlantic. Perhaps, we reasoned, the boat had simply sailed across a geographical boundary of some kind, a predictable interface of temperature or chemistry, and had thereby entered the true northern ocean.

Whatever the cause of this turquoise sea, the effect appeared to be widespread. With a strong southwest wind behind her, the sailboat was surfing down large following seas at an average speed of nearly seven knots. At this rate she was logging something on the order of one hundred sixty miles a day. And the turquoise sea persisted... for one day... for two days... for nearly three days (or better than four hundred fifty sea miles), before the sailboat suddenly emerged from the milky-blue world and into a gray-green ocean once again.

As so often happens when one is presented with a set of phenomena that do not seem to fit our standard definitions for reality, my crew and I soon found ourselves doubting our own senses, discounting the strange experience we had just shared, forcing it back in our minds until it was only a vague memory. As soon as the sea turned green again, we simply stopped thinking about the three days we had spent in that turquoise world. We allowed the experience to retreat into a convenient amnesia and did not talk about it again.

Nearly a full decade passed before the subject resurfaced for me. I was reading at the time about some of the cutting-edge science that was beginning to emerge as a result of the data being gathered by Earth-observing satellites. One of the earliest such satellites carried an instrument known as the Coastal Zone Color Scanner, or CZCS, a digital imaging device that allowed the generation of sequences of images of the Earth's surface as seen from space. The CZCS was launched by NASA in late October 1978 and was operative for nearly eight years, until late June 1986. During these eight years, it broadcast a steady stream of data back to scientists on the ground, providing them with the first record of sea surface color, radiance, and reflectivity ever to be assembled on a global scale.

Among the researchers working with these data, one in particular caught my attention, an oceanographer, then working out of the University of Miami, by the name of Dr. William "Barney" Balch. Ever since the early 1980s, Balch had been intrigued by a series of color features that had been showing up on the data record of the CZCS in certain regions of the world ocean—particularly in the North Atlantic and North Pacific Oceans, the Gulf of Maine, and the shelf and slope waters east and south of Argentina. These features, which had a unique creamy blue and white color, stood out dramatically from the rest of the ocean surface. During the eight years of data provided by the CZCS, the most notable examples were almost always found in the sub-polar North Atlantic, usually south of Iceland and south and west of the Faroe Isles. The largest of these, observed for the first time by oceanographer Patrick Holligan in the spring and early summer of 1983, was described as having an area exceeding a quarter-million square kilometers (about 96,000 square miles), or roughly the size of the state of Florida.

Once having received even an inkling that there might be a scientific explanation for the strange phenomenon my shipmates and I had witnessed, I was eager to learn more. A telephone call to Barney Balch's laboratory at the University of Miami was rewarded by a lengthy conversation with his research assistant, Catherine Kilpatrick, and the beginnings, at least, of an explanation of the ocean features Balch was then studying.

The conditions known by the popular term "milk sea," Kilpatrick explained, are the result of massive blooms of microscopic phytoplankton, called coccolithophorae algae, that occur from time to time in the surface layer of the open ocean. Early in their life cycle, these algae develop a shell-like armor of tiny round calcite platelets or "coccoliths" that together serve as a protective covering for the organism. In a later phase of their development, the algae shed the platelets in dense concentrations into the surrounding sea water, where they remain in suspension for a period lasting from several weeks to several months, creating the creamy turquoise color and high reflectivity of the ocean surface. Finally, after the bloom has subsided, the platelets slowly settle to the sea floor, where they collect in layers over eons of time to form the calcite strata that one can see uplifted in the white "chalk" cliffs of southern England, the southwestern Baltic, eastern Indonesia, and dozens of other places around the planet.

Even before the advent of Earth-observing satellites, these blooms were not unknown to science, for they had been reported anecdotally by mariners for generations. Without any means of measuring their size or distribution, however, (and without any verifiable proof of their massive scale), scientists were unable to deal in any disciplined way with these features, and they remained shrouded in mystery, out at the fringes of the known scientific world.

All of this suddenly changed, Kilpatrick explained, with the stream of data being delivered back to researchers by Earth-observing instruments such as the CZCS. Now within a few short years, the critical importance of open ocean algae blooms had become evident for all to see. As widespread inhabitants of almost all oceans of the world, the coccolithophorids were now understood to be serving an important function at the base of the food chain, helping to provide primary biological production (and thus to sustain ocean food webs) throughout their geographical range.

As net producers of CO_2 during the stage when they developed the calcite platelets, they were also recognized as having a significant influence over the rate and magnitude of carbon-transfer between the oceans and the atmosphere, making them an important link in scientists' evolving understanding of the ocean's role as a potential CO_2 sink.

Finally, as producers of dimethyl sulfide (DMS) throughout much of their life cycle, they were now also understood to be major players in the regional water cycle and heat budget. The DMS entered the atmosphere in the form of sulphuric aerosols, which in turn served as the "core" particles around which tiny droplets of water vapor were permitted to form. This increased water vapor, in the form of increased cloud cover, affected the regional heat budget in two ways: first, by trapping heat already present in the seawater, and second, by reflecting a portion of incoming sunlight away from the ocean and back out to space. When combined with the high reflectivity of the blooms themselves, these heat-related processes were becoming increasingly significant inputs for climatologists in their attempts to model the heat budget of the oceans (and, finally, the heat budget of the planet as a whole).

At the conclusion of the eight-year record of the CZCS, according to Kilpatrick, a number of fundamental questions remained. Were the size and frequency of the algae blooms increasing, decreasing, or remaining the same? If the blooms were increasing, was there any connection between such increase and the warmer ocean surface temperatures that were then being recorded planet-wide? If the blooms were decreasing, could such decrease be related to changes in the quantity or intensity of ultraviolet radiation reaching the ocean surface as a result of reductions in the Earth's ozone shield? And finally, regardless of how the magnitude of the blooms might be changing, could the chemical and biological processes that were driving them be responding in as yet undetected ways to the steady buildup of CO_2 in both the atmosphere and the oceans that was taking place as a result of human activity on the planet?

In the early 1990s, these were open questions that were driving an entire generation of researchers to try to learn more about the interchanges taking place in the surface layers of the open ocean. Nearly two more decades would pass, however, before enough new data became available from a series of increasingly sophisticated successors to the CZCS for scientists to begin to frame some tentative answers.

Among the next generation of Earth-observing instruments, NASA's Sea-viewing Wide Field-of-view Sensor (or SeaWiFS) was the most ambitious. Finally launched in the summer of 1997 after four years of postponements, the SeaWiFS mission was designed to provide "the most comprehensive global biological record ever assembled," enabling scientists to study the fate of carbon in the atmosphere, the length of the terrestrial growing season, and the vitality of the ocean's food web.

"With this record we [will] have more biological data... than has been collected by all previous field surveys and ship cruises," commented SeaWiFS project manager Gene Feldman. "It would take a ship steaming at six knots over four thousand years to provide the same coverage as a single global SeaWiFS image."

During the first three years of this mission, from September 1997 to August 2000, the biological record showed a steady increase in total global photosynthesis, driven, in part, by a series of coccolithophorid blooms that began to occur on an annual basis in the Bering Sea. The first of these blooms occurred in the autumn of 1997 and eventually grew to cover an area one-third the size of Alaska. It remained in place over a period of several months, persisting with such vigor that it changed the chemistry of a huge portion of the Bering Sea surface water, disrupting (among other things) the ability of salmon to find their way to their home rivers, and thereby prohibiting an entire year-class from spawning.

For the years 1998 and 1999, this Bering Sea bloom occurred in correlation with a strong El Nino in the central Pacific and a three degree (Fahrenheit) warming of the surface of the Bering Sea, leading observers to suspect once again that there might be a causal relationship between the increased algae blooms and the warmer sea surface temperatures. In the third year of the SeaWiFS record, however, even as the magnitude of the blooms continued to increase, the sea surface temperatures began dropping back toward more normal values. "With three years of observations we can see seasonal changes in plant and algae chlorophyll levels very well," commented NASA oceanographer Michael Behrenfeld, "but we don't yet have a long enough record to distinguish multi-year cycles like El Nino from fundamental long-term changes caused by such things as higher carbon dioxide levels in the atmosphere."

The longer record that Berhenfeld and his colleagues needed in order to understand the dynamics of ocean algae blooms was not going to come from a single instrument or a single Earth-observing mission. Instead, it would have to be compiled from a series of instruments and a lengthy sequence of such missions, each one contributing additional years of data to the record.

The prospect of linking the data from successive missions was not as simple as it sounded, for in order to combine the measurements into a continuous stream, scientists needed a base-line in real time and a method of calibration that could be repeated from instrument to instrument. No such method had yet been devised during the eight years of the CZCS, which meant that all those data, as important as they had been at the time, could not be checked for accuracy against later records and could never be correlated in any useful way with the data coming from SeaWiFS and later instruments.

Such a disconnect could not be permitted to happen again—this much was obvious to everyone involved. Calibration thus became a critical component of the design of SeaWiFS and all later missions. For SeaWiFS, a deep ocean buoy called MOBY was installed near Hawaii to make direct measurements of chlorophyll in the surface layer of the ocean. These measurements were then compared in real time with the measurements coming back from the satellite, and the calibration of the latter was adjusted.

Sailing in a Milk Sea near the Hardinger fiord, western Norway

In addition, a team of NASA scientists was placed aboard an English supply ship making regular provisioning voyages between London and Antarctica. The ship, traveling through 110 degrees of latitude on each leg of its journey, provided the perfect platform for collecting daily measurements of calcium carbonate, chlorophyll, dimethyl sulfide, and various other chemical and biological variables in the surface water, then comparing them to the orbital data for calibration.

In May 2002, with the launch of NASA's Earth Observing System's AQUA platform, SeaWiFS was joined in orbit by yet another surface-imaging instrument called the Moderate Resolution Imaging Spectroradiometer or MODIS. Meanwhile, the Japanese, the French, and several other nationalities had also started launching their own Earth-observing missions, each with its particular variety of sea-viewing instrument. With the problems of calibration now at least partially resolved, data from these new sea-surface imagers could be combined into a common pool, extending the data stream at least through the end of the current decade and providing scientists with the uninterrupted set of measurements they had been waiting for as they attempted to model the basic chemistry and biology of the ocean surface.

The story of the coccolithophorid blooms, like the blooms themselves, would seem to have no end—or at least no end that we can presently predict. The milk sea continues to erupt in gigantic explosions all around the world ocean, just as it has for countless millions of years. The uninterrupted record from space, now barely a decade long, documents only a tiny fraction of this process. If there are long-term changes taking place, they are still too subtle to detect amid the "noise" of normal year to year variability.

The story does have one additional chapter, however—a somewhat alarming chapter whose subject may turn out to be of critical importance to the health and well-being of marine eco-systems throughout the world-ocean. The chapter begins with calcium carbonate ($CaCO_3$), the basic building block that the cocolithophorids—along with thousands of other shell-building organisms—use to create the calcite armor that protects them during a critical phase of their life cycle. Nobody knows exactly how these shell-building organisms produce the calcium carbonate they need, for it does not precipitate out of sea water as a solid. To create their chalky coverings, the organisms must somehow manipulate several chemical substances in the water column.

Under normal conditions, the chemical composition of the ocean's surface provides an ideal crucible for this shell-building activity. But ever since the beginning of the industrial revolution and the large-scale burning of fossil fuels by human society, the chemical composition of the ocean's surface has been changing. Over the past two hundred years, human activities have added over two hundred fifty billion tons of carbon dioxide to the atmosphere, nearly half of which has been absorbed into the surface layers of the ocean, resulting in a chemical process that scientists have come to term "ocean acidification."

Oceanographer Ulf Reibesell explains the process this way: "When carbon dioxide dissolves in seawater it forms carbonic acid. Part of [this acid] is neutralized by the carbonate buffer, a chemical reaction that consumes carbonate ions—the building material used by calcifying organisms to produce their shells and skeletons. The remaining carbonic acid leads to a decrease in the pH of seawater [and thus to an ocean of increasing acidity]."

Besides the acidification of the ocean surface—a process that has already caused a significant decrease in the water's pH over pre-industrial levels—this drawdown of the available supply of carbonate ions means that the calcifying organisms may be finding it increasingly difficult to engage in their shell-building activities. Reibesell, along with a group of colleagues from the Wegener Institute in Bremerhaven, Germany, ran an experiment several years ago in which sample populations of two species of coccolithophorids were cultured in seawater tanks that were exposed to increasing concentrations of atmospheric CO_2. The results of the experiment were as worrisome as they were predictable: as CO_2 levels rose, the shell-building organisms became less and less capable of extracting calcium carbonate from the seawater, leading to "reduced calcite production" accompanied by "an increased proportion of malformed coccoliths and incomplete coccospheres."

As the percentage of CO_2 in the atmosphere continues to rise in coming decades and as the pH of the ocean surface continues to fall, one can only assume that the shell-building organisms will face increasing stresses. "There is a high risk that many calcifying groups may lose their competitive fitness," Riebesell observes, leading to "the loss of marine biodiversity with presently unforeseen consequences for marine ecosystems and food webs." In a more pessimistic moment, reflecting on a future scenario involving the wholesale collapse of plankton-based ecosystems, he puts it this way: "The risk is that at the end we will have the rise of slime."

In the years since *Brendan's Isle* and her crew and I first sailed through the milk sea, the percentage of carbon dioxide in the Earth's atmosphere has risen from approximately 355 to more than 380 parts per million (ppm). The rate of increase, which slowed in response to conservation efforts during the 1990s, has now rebounded to something over 1.5 ppm per year, the highest annual rate of increase since record-keeping first began in the early 1950s.

During all this time, *Brendan's Isle* and I have encountered blooms of coccolithophorids on three subsequent occasions. The first was in the summer of 1987 in the Gulf of Maine on a passage from Cape Cod to eastern Nova Scotia. The second was in the late spring of 2001, just seaward of the Hardinger fjord on the western coast of Norway. The third was in the Gulf of Maine again, during a voyage to Labrador in the summer of 2004.

Unlike our original experience in the North Atlantic, each of these later encounters has come to feel familiar, like a long-awaited reunion with an old friend. No longer strange or confusing or other-worldly, the turquoise sea now presents itself as exactly what it is: a self replicating life force, a massive colony of living, breathing protoplasm, a fundamental building block upon which life in the oceans has evolved; perhaps (who knows?) the benign countenance of life itself.

As scientific knowledge about the coccolithophorids and their role within ocean eco-systems has increased, and as threats to the health and evolutionary fitness of these organisms have become more clearly understood, I've grown increasingly concerned each time I've sailed into their milky realm. Could ours be the last generation of sailors to watch the sea blossom with life as it has for half a billion years? Will our children have an opportunity to witness, as we have done, the eruption of massive primal food sources large enough to fuel an entire marine biota? Will our grandchildren still know an ocean teeming with complex hierarchies of algal-based life, or will they be left to witness a kind of evolution-in-reverse in which the jellyfish reign supreme?

I ask these questions, then reluctantly set them aside, for only time will tell whether our newfound knowledge will provide the incentive for human societies to make the massive reductions in global carbon emissions necessary to change the story of the cocolithophorids. Meanwhile, as the sailboat slips through an ocean saturated with trillions of tiny calcite mirrors, I look out at the iridescent surface and understand at last that I am surrounded by a huge living presence. I close my eyes, the better to feel the boat rising and falling on the turquoise swell, and I imagine myself a participant in the breathing of the biosphere.

It's just sand…and it moves as though it were itself a slow and ponderous schooner on the sea, seeming to breathe as schooners do, rising and falling on the very breast of the planet… but it could be the shape-changer and face-dancer of aboriginal legend, yielding in order not to yield.

Marq de Villiers and Sheila Hirtle, *Sable Island*

15 A Dune Adrift in the Atlantic

A HUNDRED SEA MILES SOUTH of mainland Nova Scotia lies a mysterious, twenty-five mile long crescent of sand that sits alone at the edge of the continental shelf—a sandbar, rising precipitously from the ocean floor, that mariners have known for the past five centuries as Sable Island.

Sable is a geological anomaly, for there is no bedrock here—no buttress of any sort to anchor the shifting sand. Instead, as core samples from a failed exploratory oil well drilled in the late 1960s have revealed, the island rests upon a thousand feet of modern (post glacial) sediments of sand and fine gravel, which in turn rest upon more than fifteen thousand feet of similar and ever more ancient sediments, collected on the ocean bed here over the past half-billion years.

The question, of course, is how did all this sand get here? What forces combined to carry the tiny grains across a hundred miles of open ocean and deposit them at the seaward edge of the continental shelf, forty miles from the great Atlantic abyss? In light of the ferocious storms that periodically sweep across the island, what geophysical mortar binds the two massive beaches and the great windswept dunes and keeps them in place century after century? Why aren't these fragile and ever-moving heaps of sand simply scattered to the wind or washed away into the sea?

These are questions I'd been asking myself ever since I first learned about the existence of Sable Island—questions I was still asking one winter several years ago as I began putting together a plan for an early August visit to Sable aboard *Brendan's Isle*. During the following months, I read about the history of the island since its European discovery in the sixteenth century by a sequence of Portuguese explorers. I learned about the more than 500 documented shipwrecks that have been littered across the beaches and shallow sands of the Sable Bank over the ensuing centuries—a fearsome toll of tragedy and death that has earned this place its well-deserved epithet, "Graveyard of the North Atlantic." I read about the herds of feral horses— some three hundred seventy-five at latest count—that roamed the dunes and beaches, their tangled manes flying in the summer wind. I met—via the written word at least—the only two permanent human inhabitants of the island: Gerry Forbes, general handyman, chief meteorologist, resident agent for the Canadian Coast Guard, and government-appointed manager of the island station; and Zoe Lucas, self-appointed steward of the Sable horses, Webmaster for the Sable Island Web site, and Research Associate for the Nova Scotia Museum, in charge of a wide range of conservation and environmental studies around the island.

The difficult part of visiting Sable Island is not just transiting the hundred or so offshore miles required to get there from the Nova Scotia coast. During settled summer weather, any experienced sailor with a well-found vessel can expect to make such a passage in less than twenty-four hours. No, the difficult part actually begins many days before, as you set about collecting information on future weather systems (depressions, fronts, precipitation, visibility, wind and wind-driven waves, long-period swell from previous systems) that will affect the "target" area during the time that you and your vessel will actually be visiting the island.

Such preparation is critical, for the long, sweeping shoreline of Sable Island contains no safe harbor. Landing on the island thus involves anchoring your vessel several hundred yards from the shore, transiting the surf zone in your rubber dinghy, and landing fully exposed to whatever wave-action may be present on the open beach. For most small landing craft, this maneuver is possible only on the north side of the island, and only after a lengthy period of settled southerly weather, when the curved, north-facing shoreline provides adequate lee from both wind-driven waves and swell. In addition, if you also want to actually *see* the island when you get there, you need to pick a day for your visit when the sun is shining and the air is relatively dry—no mean trick in a region of colliding air masses that is situated smack in between the warm surface waters of the Gulf Stream and the frigid remnants of the Labrador Current. Boasting an average of more than 125 days per year with reported visibilities of one kilometer or less, Sable Island is considered by meteorologists to be one of the foggiest locations on earth.

In the days before the journey was set to begin, Kay and I waited poised and ready in Cape Breton's Bras D'Or Lakes, just a few miles from our planned point of departure on the Nova Scotia mainland. During the final week of July, I started monitoring evolving weather patterns, hoping to find an appropriate weather window for our passage. On July 31, two old friends arrived by air from the States: Amanda Lake, veteran of many thousands of sea miles with us over the years, and John Griffiths, master carpenter and builder of *Brendan's Isle*. The next day, with a ridge of high pressure moving in from the west, we made the passage down the Lakes and out onto the Atlantic coast to the little harbor of D'Escousse. Here we met with another old friend, Silver Donald Cameron, a well-known Canadian author and journalist who, despite his many years of living and sailing in the Cape Breton area, had also never made the journey to the sands of Sable Island.

The gods must have been smiling on *Brendan's Isle* on the morning of August 2, for the dawn broke crisp and cool, with clear horizons all around and a light southeasterly breeze. After one more check on the forecast, I and my shipmates retrieved *Brendan's* anchor and pointed her bows due south, directly toward that mysterious mid-ocean sandbar that had fired my imagination for so long.

The actual passage to the island was largely uneventful. Shearwaters soared across the gently undulating swell; Mother Cory's chicks fished all day in the sailboat's wake. In the afternoon a small family group of minke whales crossed ahead; just before sunset a troupe of porpoise interrupted their travels to play in the eddies beneath the cutwater. Later, with the decks illuminated by a waning gibbous moon, a bright light appeared on the southern horizon—a safety flare (we learned later) from the Thebaud gas platform—followed by the tiny flash of a lighthouse far away to the east. Don and I were on deck for the sunrise watch, so we shared the experience of seeing Sable Island loom up out of the gray pre-dawn light, a pencil-thin line slowly growing more distinct all along the rim of the horizon.

The protocol for making a landing on Sable includes the acquisition of a letter of permission from the Canadian Coast Guard, followed by a series of telephone calls to island manager, Gerry Forbes, for final clearance and specific instructions on where to anchor and come ashore. In our particular case, with a light southerly breeze and nearly flat seas, the process of landing in our 11-foot Zodiac was dead easy. After making one last phone call, we just waited for Forbes to arrive in his black four-wheel drive pickup truck and gesture to the spot on the beach where he wanted us to disembark.

"Welcome to Sable Island," intoned the island manager as we carried our dinghy safely above the high-water mark on the beach. "You've chosen one of the most beautiful days of the summer for your visit."

As my shipmates scurried across the sand and tumbled pell mell into the extended cab of his pickup truck for the drive back to the island station, I took a moment to observe our host more closely. My first impression was of a somewhat raggedy master of a little ship—perhaps a rescue vessel of some kind—parked 100 miles out in the Atlantic and waiting for the next crew of wayward sailors to come to grief upon the long, featureless sands of the Sable Bank. He was dressed mostly in black, with a loose-fitting black tunic, matching trousers, and heavy black walking boots. Above a pair of wrap-around sun glasses, he wore a thatch of fly-away gray hair and a dark blue brim cap, decorated with a gold Sable Island logo. As he welcomed us to the island, I quickly realized that he was indeed the commander of this "ship" and that we were here to enjoy its wild beauty entirely at his invitation and convenience.

As we made our way across the soft, deeply rutted sand toward the main station, Forbes entertained us with a constant stream of information about the island, the beaches and dunes, the flora and fauna that inhabit this place. "You'll see the wild horses everywhere," he said. "You'll meet a different family group behind every sand dune. They're entirely fearless—

You'll see the wild horses everywhere.

but you're not permitted to touch them or to interact with them in any way. If you do, Zoe will be very upset. The rule here is nature's rule. We are—all of us—pledged not to interfere."

No sooner had we clambered down from his truck near a cluster of white frame buildings than the talkative manager was off on another errand. An airplane was scheduled to land on the south beach in less than an hour, he explained, and he and several helpers needed to mark out a runway in the sand.

"But here comes Zoe," he said, gesturing toward a small, dark figure walking briskly toward us. "She'll be able to answer any other questions you may have. Then if you'd like, you might want to wander on your own out to the west end. I'll look for you back here at the guest barrack after lunch."

Next moment we found ourselves standing in a semi-circle, introducing each one in turn to Zoe Lucas, the only other full-time resident of Sable Island. Like Forbes, Zoe lived and worked on the island all but a few weeks a year (as she had done for more than thirty years). She was short and petite, with a pixie-like appearance, dark brown eyes, small mouth and nose, a fleeting and somewhat ambiguous smile. She wore a loose-fitting brown jacket, gray baggy trousers, a pair of well-worn sandals. Her hair was long and unruly, bleached in the sun to a dozen shades of blond and brown and gray, not unlike the mane of one of her beloved Sable horses. She wasted no time with polite banter. After finishing with introductions, she turned to more serious concerns. She talked about the island, the horses, her work.

"I came here in the 1970s to take part in a major attempt at dune restoration. You can still see the remains of the wooden buttresses and snow fencing that we used to try to stop the moving sand. It took us ten years to realize that all our efforts were in vain. Nature rules here—in everything. The best we can do as human beings is to get out of the way, stop trying to control things, let the natural rhythms take over."

I asked her about the island, about its future. I had read that some scientists were speculating that the sands here might be slowly migrating to the east and disappearing back into the sea. With the abandonment of the dune restoration project, what were the chances that Sable Island would still be here in the foreseeable future?

"Nobody knows exactly what combination of forces have combined to create this island—or to keep it where it is. And nobody knows for certain what's going to happen next. What we do know is that the island exists at least partly because of a complex combination of ocean currents that gyre around this particular part of the Atlantic. They've been turning like a huge slow centrifuge with Sable at the center—and Sable's sands have been collecting here, precipitating out of the gyre, and settling to the ocean floor for tens of thousands of years.

"The danger," Zoe added, "is not from mother nature. The danger is from us. From the oil and gas drilling, the manipulation of the dunes, the attempts to "stabilize" processes that are inherently dynamic. Sable will be around for many more thousands of years, I'm certain, as long as we're able simply to let it be."

The sun climbed steadily in the southern sky and the day warmed quickly as we took our leave of Zoe and headed out on foot along "Highway One" (as the only road on the island was called) toward the fresh-water lakes and the abandoned lighthouse at the island's west end. We tried to be careful not to startle the foraging groups of feral horses or to disturb them with our presence—although to be honest, the animals seemed to take no notice at all of their human visitors. (Zoe's policy of non-interference with these animals certainly seemed to be working, I mused, for they behaved toward us with utter indifference, as if we were invisible.)

In the afternoon, after finishing our bag lunches back at the guest barrack, we were treated to an even more remarkable tour of the eastern portion of the island, this time courtesy of Gerry Forbes and his four-wheel drive pickup truck. Even a cursory glance at the scale of the massive south-facing beach (dozens of miles long and, in places, more than half a mile wide) made it obvious why we weren't attempting to traverse this portion of the island on foot. After a brief visit to the aircraft and a check on the temporary runway for sink-holes, Forbes took us on a fast ride along the surf-line to visit with the vast herds of gray and harbor seals that gathered here every year to whelp and raise their pups.

Farther down the beach, we stopped to inspect the site of the most recently documented shipwreck on the sands of Sable Island—a forty-one foot fiberglass sailing yacht named *Merrimac*, driven ashore on a still, foggy night in July, 1999, broken up on the beach, and swallowed days later by the all-consuming sand. All that remained of the hapless little ship was a plywood bulkhead, bleached and splintered, that had been disgorged from the sand during a recent storm.

The final moments of our visit took place once again at the invitation of the island manager, after we had returned with him to the main island station. "Before you leave, I must show you something of our most important

The remains of the Merrimac, *driven ashore in 1999*

mission here," he said. "In these larger buildings you'll find all the various apparatus for the island's meteorological station. In the smaller buildings are the instrument packages for a variety of important climate and atmospheric studies sponsored by Environment Canada, NASA, Princeton University, and a number of other academic and scientific organizations."

With this preamble, Forbes began leading us through a rather daunting array of scientific devices, pausing at each one to explain its operation and purpose. The most interesting to me were the instruments designed to monitor the chemistry of the Earth's near-surface atmosphere, for here, on an island that seemed so remote and pristine, were instruments measuring ever-increasing levels of industrial aerosols, chemical insecticides and herbicides, methane, CO_2, and other human-generated carcinogens, pollutants, and greenhouse gasses. "Because of its location at the edge of the continent, Sable is in a unique location for making such measurements," Forbes explained. "When the winds are easterly, we measure only traces of many of these substances. But when they blow from the west… well, then many of the measurements go off the scale and Sable might as well be at the end of a huge continental sewer pipe."

With these rather sobering observations, our host concluded his re-marks. Soon we were collecting our knapsacks and carry-bags and scam-pering into his pickup truck once again for the two-mile drive back out to the north beach and the dinghy ride to the anchored sailboat. Standing in a small cluster on the beach, we tried to express our gratitude for the day that we had just been given. Yet as sincere and heartfelt as our thanks obvi-ously were, they were also oddly subdued—perhaps because my shipmates and I were still pondering the bundle of contradictions that Sable seemed to represent.

Here was an island administered according to "nature's rule," whose two human stewards were pledged to keeping human interference to a minimum and letting the natural rhythms predominate. Yet here also was an island vulnerable in the extreme to the "human interference" of society at large—to the habits and excesses of millions of human beings who had never seen nor heard of this place. What would happen to the sands of Sable, I wondered, if the steady increases in atmospheric CO_2, as measured both here and worldwide, were to result in accelerating glacial melt and subsequent sea-level rise, as many scientists were now predict-ing? What would happen if changing patterns of global climate were to result in the alteration of surface and sub-surface circulation in the world ocean, changing the behavior of "permanent" ocean features such as the Gulf Stream and the Labrador Current? Would the massive gyre that Zoe had described continue to circulate around the island? Would the sands continue to precipitate out of suspension and replenish the beaches and dunes? Or would the entire complex process that we know as Sable Island simply cease to function and disappear into the sea?

As we concluded our goodbyes and pulled away from the shore, I gazed for one last time at the long crescent of beach, and I found myself hoping against hope that this starkly beautiful island of sand could somehow sur-vive the ravages of coming decades, persisting according to "nature's rule," as wild and free as the horses that roamed its dunes. Sable Island, I now understood, was very much like the island we call Earth: an oasis, poised in exquisite equilibrium. It was a precious gift, whose ability to sustain life—and indeed to sustain itself in its present configuration—was dependent upon a complex system of geophysical forces only dimly understood by the very creatures who collectively held the power to alter and disrupt them.

I hammered on the gate, making the iron chain rattle on the dark wood. After a minute, an old man appeared. "Peace be with you," I said. "And also with you," he replied.... I followed him through the courtyard to the guest room. He put a cushion under me, gathered twigs, fed the fire, and blew on it until it flamed fiercely. He asked me if I would give him my socks so he could dry them. I did so gratefully. Then he left and returned with a pot of sugared tea and... plates of rice and spinach... Only when he saw that I was warm and had finished eating did he lean forward and ask, "And who are you? And where are you from?"

Rory Stewart, *The Places In Between*

16 The Beneficent Gene

MANY YEARS AGO, LONG BEFORE *Brendan's Isle* was even a vague notion in my mind, Kay and I and our three boys made a summer voyage to the coast of Maine aboard an old wooden sloop called *Westral*. The weather was particularly bad that summer. Day after day the rain poured down in buckets and the fog hung about the headlands thick as night. What's more, the boat developed a seemingly endless series of mechanical problems—a combination of circumstances that kept us harbor-bound for days on end.

Cabin fever was fast becoming epidemic. The children needed to get off the boat and do something ashore...but with the cold and fog and rain, what could we do that wouldn't result in a generalized encounter with hypothermia?

"Rent a car," suggested Kay. "Drive down to Portland and let them run around in a mall. Take them to a movie."

The idea seemed sound except for one serious drawback, which we learned an hour later on our visit to the only automotive service station in town. "We don't rent cars," said a voice that arose from somewhere underneath the rear axel of an old Chevy pickup truck. The voice was soon joined by the legs and torso and head of an elderly fellow wearing a blue woolen watch cap, an Abe Lincoln beard, and brown, grease-stained coveralls. He gathered himself up from the floor, wiped his hands ceremoniously on an oily rag, and peered pensively from face to face at the five rain-soaked sailors standing before him. "We don't *rent* cars," he repeated, "...However, we do sometimes *lend* a car to folks who seems to need one."

Without another word, he stepped over to a desk in the corner of the shop, picked up a set of keys, and handed them to me. "It's the green Ford station wagon out in the front lot. She's all fueled up. If I'm not here when you get back, just slip the keys under the door."

We never exchanged a dollar—never even exchanged names with our grease-covered benefactor. We just drove off in his car, heading god-knows-where, leaving him with nothing except a chorus of thank you's and a promise that we'd try to be back before dark.

During that soggy Maine afternoon all those years ago, such an unpro-voked act of generosity and trust seemed difficult to comprehend. At the time, I still hadn't been sailing very long or far, but I nevertheless thought I had a reasonable understanding of the way the world worked. Something had to be amiss, I decided. Normal people just didn't hand over the keys to their cars and let you drive away without even asking your name.

Now, forty years and tens of thousands of sailing miles later, the mem-ory of that old mechanic's kindness doesn't seem nearly so difficult to un-derstand. The altruistic impulses that stirred his heart during his encoun-ter with the five rain-soaked strangers at his door are the same impulses that we have witnessed over and over again in chance encounters with people everywhere we have sailed. From Maine to Martinique, from Nova Scotia to the fiords of Norway, my shipmates and I have been the unwitting recipients of an astonishing outpouring of hospitality and human kind-ness. The encounters have taken many forms. Sometimes they've centered around the gift of a fish, or the offer of a dinghy ride, or the loan of a car. Other times they've started with the invitation to come ashore to take a hot shower, to share a cup of coffee, or to join with family members for a meal and an evening of story telling.

One encounter that I remember with particular clarity took place a few years ago on the western coast of Norway. Kay and I, along with our friend Cherie, found ourselves sailing quite by chance one afternoon into a little cul-de-sac cove called Dragvagen, just a few miles from the city of Molde, where we had promised to rendezvous in a few more days with friends who would be joining our crew. The cove we found was small and quiet, sur-rounded by farms and summer cottages. As soon as the anchor was down, Cherie took the rubber boat ashore on a scouting party. A short time later she returned with a young man named Tore and his wife Sati, owners of one of the small farms that bordered the cove. By the time they'd climbed aboard *Brendan's Isle*, the couple had already offered showers for the crew, a machine to wash our clothes, bicycles, a place to moor our dinghy, a ride to the airport, a place to park a rented car. As we talked, they also invited us to a barbecue dinner at their home the next evening to meet their family and neighbors.

Because we had several days to wait before our new crew was due to arrive, we had time to savor the many kindnesses extended to us by our Norwegian hosts. Next day after breakfast, I accepted Tore's invitation to drive me to the airport to pick up a rental car. In the afternoon, he and Sati took us into the city for a few hours of live music at a local jazz festival, then treated us as guests of honor that evening at a huge community get-together at their farm.

By the time our new crew members were due to arrive thirty-six hours later, I was finding it difficult even thinking about the moment when Kay and Cherie and I would need to bid farewell to Tore and Sati. As I pulled out of the farm lane on my way to the airport, I gazed back at the solitary sail-boat riding at anchor a hundred yards from the shore, at the sturdy wooden pier where our dinghy lay moored, at the steep-roofed farmhouse where we had spent so many pleasant hours. Tonight, I realized, there would be embraces, a few tears, and promises that somehow, some day, we would find a way to get together again. Then, knowing that our promises would not be kept, we would sail away, filled with gratitude, and wondering how we could ever repay all the kindnesses that had been showered upon us.

Perhaps because the human species is so prone to violence, aggression, and all manner of generalized mendacity, it is difficult, when thinking about our customary patterns of social interaction, to understand such unpro-voked acts of charity toward total strangers. In the first few years of our cruising, as I've said, Kay and I tended to think of each such encounter as somehow standing apart from general norms of human behavior. The longer and farther we sailed, however, the less satisfactory this explanation became. There were too many generous people everywhere we traveled to imagine them all as exceptions. There was the old postmaster in the Newfoundland outport who invited us to take off our shoes at his door and enter his house any time "like we was family." There was the North Carolina cotton farmer who drove us twenty-five miles in his pickup truck to the nearest grocery store (and twenty-five miles home again). There was the Labrador fisherman who wouldn't be satisfied until we had accepted, as a gift, the fattest and most beautiful cod in his dory. There was the Icelander who stopped his work for a day, loaded our entire crew into his car, and drove us into the mountainous interior of his country in order to share with us some of the places he loved. There was the old Scot who helped to pilot our sailboat past a series of unmarked dangers and into the anchorage in front of his house and who later graced us with a tub of haggis and a jug of his best Scotch whiskey.

A pattern of sorts was emerging here, as we eventually came to realize—an outpouring of generosity and good will, freely offered to unknown passers-by, without any requirement for recompense or reward. What's more, we were hardly the only recipients of such benevolence. Other sailors—especially those who had traveled to remote and sparsely populated coasts—told us of their own similar experiences. And other travelers who were not sailors (but who traveled by similarly modest conveyances) added their own testimony, describing the kindnesses afforded to the walking man, the bicycler, the kayaker, the wagoner, the horseman, and others of peaceful intent who took to the open road.

One such non-sailor with an astonishing story to tell is Rory Stewart, a young Scotsman who set out in the late autumn of 2002 to walk across the winter mountains of Afghanistan, from Herat to Kabul. In his book *The Places in Between*, Stewart describes his harrowing five-week ordeal, crossing an impoverished land of tiny fiefdoms and petty tribal chieftains, traveling without any means of self-defense except his own wits and his trust in the fundamental goodness of the people he would encounter along the way.

Stewart was no stranger to this kind of travel, for during the sixteen months prior to his trek across Afghanistan, he had walked nearly the entire way from western Iran to the foothills of the Himalayas, crossing a large portion of northwestern Asia without any assurance from one day to the next that he would find food or shelter or protection from those who might harm him.

> *I was alone..., walking in very remote areas; I represented a culture that many of [the local people] hated, and I was carrying enough money to save or at least transform their lives. In more than five hundred village houses I was indulged, fed, nursed, and protected by people poorer, hungrier, sicker, and more vulnerable than me. Almost every group I met—Suni Kurds, Shia Hazara, Punjabi Christians, Sikhs, Brahmins of Kedarnath, Garhwal Dalits,*

> *and Newari Buddhists—gave me hospitality without any*
> *thought of reward. I owe this journey and my life to them.*

Originally denied a visa for entry into Afghanistan, Stewart detoured around that war-torn country in order to continue his journey in lands farther east. Then, a few weeks after the American routing and temporary expulsion of the Taliban, he managed (without a visa) to slip across the Iranian/Afghani border into Harat. Here he pleaded with local officials for permission to set out on his journey. "It is mid-winter," one of the officials told him. "There are three meters of snow on the high passes, there are wolves, and this is a war. You will die, I can guarantee. Do you want to die?"

Confident of the ancient traditions of hospitality among ordinary Afghanis, Stewart ignored the official's warning and set off on a little-traveled footpath into the foothills of the Paropamisus Mountains. The farther he traveled away from the major thorofares, it seemed, the more generous his benefactors. In nearly every mountain village he visited, the feudal chief would offer him food and shelter, often in the guest quarters of his own dwelling. Then, next day, the chief would provide a letter of introduction to his counterpart in the next village and an escort to show the way. "I was passed like a parcel down the line," writes Stewart, "from one chief to the next. Their men were willing to walk a full day through the snow to accompany me and a full day back. I always insisted they take some money, but they were doing it as a courtesy for me as a traveler...."

Meanwhile, on the other side of the world, another traveler was making a journey similar in many ways to Rory Stewart's Afghanistan adventure. The traveler in this case was an American horse-breeder and trainer, Dave McWethy; the means of travel was a homemade gypsy wagon and a trio of sturdy Norwegian fiord horses; the setting was a ribbon of rural highway stretching across the entire width of the northern United States, from eastern New York State to western Montana.

For McWethy, the reasons for making such a journey were clear. It would serve as a powerful means of communing with his beloved fiord horses. It would provide a dramatic way of encountering the landscapes of rural North America. And who knows, risky as it was certain to be, it might also serve as a means of connecting with ordinary Americans in ways that most people are seldom able to do.

Early in his diaries of this journey, McWethy admits of his anxiousness over the uncertainties that lay ahead. Where would he stop each night? How would he stable his horses? Where would he find food for himself and hay for his animals? How would he (and the animals) survive the snow and cold of a long northern winter and the blistering heat of a high plains summer?

He assumed from the beginning that he would sleep and take his meals in the gypsy wagon, purchase food in small towns along the way, and pay for stabling for his animals when the weather turned inclement. Little did he realize that every day for the next eleven months, every one of these basic needs would be provided for him. Day after day, evening after evening, he became the recipient of an outpouring of down-home hospitality the likes of which he had never dared to imagine. When he was near a town, he was inevitably offered stabling and forage for his animals in a public park or a local fairground, then invited for dinner with the family of a prominent citizen or with the local Lions or Kiwanis chapter, often in exchange for an impromptu talk about his journey. When he was in the countryside, he always managed to find a farmer ready to welcome him and help with his animals. Sometimes he would simply be offered a corner of the farmyard to park his wagon and paddock his horses. More often, however, he would be welcomed to dine as a guest with the farmer and his family (and to spend the night under their roof if the weather was stormy), while his horses were bedded in straw in a warm and dry barn, and fed with the farmer's own alfalfa hay.

"Almost everywhere we stopped," writes McWethy, "hay was provided for us. At first I did not understand, and I offered to pay. But I always got a response like my offer was a breach of etiquette. For just like the food and shelter given to me, the hay, too, was a gift of hospitality."

By the end of his journey, McWethy had discovered the lesson that travelers all over the world seem continually to discover: that there is a powerful inclination among ordinary human beings to perform acts of kindness and charity toward the stranger who appears at their door, especially when such stranger is perceived to be of peaceful intent, vulnerable to the elements, in need of food or shelter, or far away from home. The more needy the visitor and the more isolated the setting, the more this charitable impulse is likely to become manifest—to the point where a traveler of simple means might conceivably cross a continent or even circle the globe dependant upon nothing other than the collective kindnesses of an altruistic humanity.

The traditional roots of this altruism go back to the beginnings of human history. The ancient Egyptian "Story of Sinuhe" recounts a time nearly four thousand years ago when this young attendant to the Pharaoh's court was forced to flee into the desert as a political refugee. He traveled into Syria, where he was treated with great courtesy and respect. In turn, when he married and established a household there, he determined to treat all travelers with the same respect: "I let every one of them stay with me; I gave water to the thirsty; I showed the way to him who had strayed; I rescued him who had been robbed...."

Meanwhile in nearby Palestine, another famous householder was treating the strangers at his door with even more lavish hospitality. The householder was Abraham, father of the three great religious traditions of the west; the occasion was a visitation by three unknown travelers at the portal of Abraham's tent that eventually became the paradigm for ancient Israelite and early Christian hospitality.

> *When he [Abraham] saw them*
> *He ran from the tent door to meet them*
> *And bowed himself to the earth and said...*
> *Let a little water be brought, and wash your feet*
> *And rest yourselves...*
> *Then he took curds and milk,*
> *And the calf which he had prepared,*
> *And he set it before them; and he stood by them*
> *Under the tree while they ate.*

Little did Abraham know that the three travelers were in fact messengers sent to him by God—perhaps the reason for the warning found later in the Old Testament: "Do not neglect to show hospitality to strangers, for thereby some have entertained angels unawares."

Elsewhere in the ancient Mediterranean world, the wanderers in Homer's Greece were also afforded a special kind of hospitality (Greek: *philoxenia*), owing to their special status as emissaries from Zeus. Thus in the *Odyssey*, the virtuous Nausicaa implores her attendants to care for the "unlucky wanderer,"

> *...For a kindness that may seem small to us is not so to him.*
> *So come, my ladies, find food and drink for our guest,*
> *And bathe him in the river, in a place sheltered from*
> *the wind.*

Beginning with these early roots, the traditions of hospitality toward the stranger have persisted through the ages and throughout the world. In the Koran, sustenance is considered a birthright rather than a gift, and the duty to supply it is a duty to Allah, not just to the stranger. In the Torah, the faithful are instructed in the importance of *tzedakah*, the divine mandate to share one's resources with those in need. In Ireland, the ancient Celtic tradition of hospitality (*oigidecht*) leads logically to the establishment of *bruideans*, free public houses of welcome, placed strategically at major intersections of roads. In Poland, the traditional Christmas eve supper (*wigilia*) includes the practice of leaving an extra plate and empty seat at the table for the stranger who might appear without warning at the door (a custom closely resembling the moment during the Jewish Passover Seder when food and wine are left at an unoccupied place at the table in hopes that the prophet Elijah, in his guise as the humble wayfarer, might enter through a doorway purposely left ajar and join the family in its ritual meal).

When considering the question of why this inclination toward hospitality is so universal, many observers have opted for the cultural argument. In this view, ordinary people the world over are reminded by their folk wisdom or instructed by their religious leaders or compelled by their societies' social mores to behave altruistically toward one another, not because such behavior is natural to the human species but because a given set of tribal or religious or cultural mores has deemed it to be so.

The problem with this kind of reasoning is that it fails to take into account the pleasure—indeed the sheer delight—that most of us feel when we are cast into the role of benefactor and allowed to help the stranger in our midst. Of course it is pleasant to be the recipient of someone else's kindness, especially when you are the one who is far away from home and in need of help from a fellow human being. But for most of us, it is always more satisfying to be the giver—for then you are transformed by your own acts into a person who is magnanimous, virtuous, and kind.

I find it hard to imagine Tore and Sati performing all the generous deeds they did for the three unknown sailors who arrived at their doorstep just because of some moral imperative they had read about in a book or had been told by a teacher or had heard from the pulpit. I find it difficult to impute similar motives to the hundreds of farmers who aided Dave McWethy on his eleven-month drive across America or to the dozens of local chieftains who fed and sheltered Rory Stewart on his five-week trek across Afghanistan.

Much more likely, it seems to me, is the proposition that every one of these benefactors was acting on an impulse that is innate and automatic—a natural propensity toward generosity that is actually built in to the human genome.

One test of such a proposition is Darwin's own: if a knee-jerk altruism toward strangers has indeed become part of our species' evolutionary blueprint, then it must have a "survival value" greater than the obvious risks that it imposes. We know that human beings are aggressive and territorial, biologically predisposed to organize themselves into small tribal units and to fight over resources. What could be the survival value, then, of opening one's territorial boundaries and sharing one's food and shelter with a wanderer from an alien tribe?

The answer, of course, has to do with reciprocity. If my tribe and I allow you to come in peace into our territory, if we feed and protect you and treat you as an honored guest, then we may justifiably expect the same treatment should one of our tribe decide to travel into your territory. In this way we all extend our ability to move more freely about without being in constant danger of attack or capture—and everybody wins.

In autumn each year after we've finished our summer sailing, Kay and I return to our farm on the Sassafras River, Maryland. Here, for the next several months, we watch a parade of cruising boats meander into the river, anchor for a day or two in one of the nearby creeks or coves, then meander out again. These are the "snowbirds," sailors from New England or Nova Scotia, Quebec or Ontario, wending their way south, down the Chesapeake Bay and Intracoastal Waterway, heading for a winter in Florida or the Bahamas or the islands of the high Caribbean.

For Kay and me, the snowbirds represent a welcome opportunity to change places in this business of take and give, for here at our very doorstep comes a group of travelers in need of the same simple acts of kindness that we have received at the hands of others year after year. Several times each autumn we venture in our dinghy out onto the river to engage the owners of an anchored sailboat in conversation. Inevitably, the results of our talking will include the mention of some small item that they may need. Then we're able to offer a shower, help with some minor repair, drive them to a grocery store, provide a meal.

One rainy autumn morning several years ago we awoke to find a little hard-chine steel cutter anchored in a cove near our house. She was rigged out for serious blue-water sailing, flying a tattered French tri-color from her transom. After breakfast, we rowed out in the rain to say hello to the four people on board: a young French man and wife and their two small children.

They had sailed across the Atlantic to eastern Canada, the man said, then down the American coast to the northern Chesapeake Bay. They were looking for a place where they could stop for a few days, rent a car, take the children ashore and dry them out a bit.

I gazed sternly at the four rain-soaked sailors standing before me. "You're on the eastern shore of Maryland," I said. "Nobody rents cars on the eastern shore of Maryland." Then I winked at Kay, trying not to smile. "Nobody *rents* cars…. However, we do sometimes *lend* a car to those that seem to need one."

Half an hour later, as Kay handed over the keys to our old Honda station wagon, I found myself enjoying the expressions of consternation and astonishment written all across the young French family's faces. They couldn't know, of course, that they'd just given us a chance to repay a kindness tendered many years before.

"She's all fueled up," I said, feeling as if I were closing an account that had long been overdue. "If we're not here when you get back, just slip the keys under the door."

Notes and Sources

Front Matter: Epigraph

p. vi *The end of all our traveling:* These thoughts of T.S. Eliot appear as they are paraphrased by John Elder, *Reading the Mountains of Home,* Cambridge (Mass), Harvard University Press, 1998, p. 232.

Introduction: Perfect Travelers

p. 13–14 *The surface of the water rises and falls:* This fanciful description of *Brendan's* landfall on the Blessed Isle of the Saints is a somewhat free-wheeling extrapolation of the description of this event as it appears in the original Latin text of the *Navigatio Sancti Brendani.*

p. 15 *[Brendan and] the monks disembarked:* Tim Severin, *The Brendan Voyage,* London, Hutchinson & Co., 1978, Appendix I, "The Navigatio," p. 250.

p. 16 *The themes…are of compassion, wonder, and respect:* Barry Lopez, *Arctic Dreams,* New York, Charles Scribner's Sons, 1986, p. 316.

p. 17 *"impeccable, generous, innocent men":* ibid.

Chapter 2: Landfall Faroe Isles

p. 41 *Kell and Mikey decide after supper:* Just a few weeks before *Brendan's Isle* was due to set sail for the Faroe Isles, Steve Arms broke his ankle and was unable to make the crossing. His last-minute replacement was an old sailing friend, "Mikey" Auth, who joined the boat in Maryland and continued all the way to Denmark. Steve, who would have to wait two more years for his first ocean crossing, remained in the States until his bone was healed, then traveled with his mother to Shetland, where they both joined the boat for the passage across the North Sea to western Norway and on to Copenhagen.

Chapter 4: The Wreck of the Braer

p. 51 *Why do we act so foolishly*: Thomas Berry, *The Dream of the Earth*, San Francisco, Sierra Club Books, 1988, Foreword, p. viii.

p. 51 *After reading the report*: On the day of the Braer disaster, the Associated Press wire service account was carried by major news media throughout North America. I first read this AP report in the *Naples* (Florida) *Daily News*, "Oil Spill Threatens Wildlife," January 6, 1993, pp. 1A and 14A.

p. 52 *The newspaper's account*: The quoted passages in this paragraph are all from the AP wire report, as cited above, p. 1A.

p. 52 *That disaster dumped thick, heavy oil*: ibid, p. 14A.

p. 54 *But now, as soon as the prevailing*: A dramatic accounting of the various species of sea life threatened by the breakup of the oil carrier was published in the London *Daily Mail*, "Helpless Victims of a Poisoned Sea," January 6, 1993, pp. 4–5.

p. 54–55 *There are many possible causes*: A discussion of the causes of the wreck and oil spill can be found in the AP wire report "Ship Operators Cutting Corners on Safety," as published in the *Miami* (Florida) *Herald*, January 6, 1993, p. 9A.

p. 55 *Shortly after the helicopter evacuation*: Details in this and the following paragraph about the rescue of crew members are from the AP wire report, "Tanker Still Intact," as published in the *Naples Daily News*, January 7, 1993, p. 5A.

p. 55 *Our first priority is the safety of life:* ibid.

p. 56 *This anthropocentrism is largely consequent*: Thomas Berry, op.cit., p. 21.

p. 56 *An accurate accounting*: For a discussion of wildlife fatalities caused by the spilled oil, see Manchester *Guardian Weekly*, "Tanker Disaster Hits Shetland," January 17, 1993, p. 3.

p. 56 *in an evocatory rather than a dominating relationship*: Thomas Berry, op.cit., p. 14.

Chapter 6: Waters of Separation

p. 65 *These are the waters of beauty and mystery*: Annie Dillard, *Pilgrim at Tinker Creek*, New York, Harper & Row, 1986, p. 266.

p. 72 *And as it darkens and the stars*: Rockwell Kent, *N by E*, New York, New York Literary Guild, 1930, p. 17

p. 73 *giant icebergs under our lee*: Desmond Holdridge, *Northern Light*, New York, The Viking Press, 1939, p. 63.

p. 73 *a grim land shrouded in scud*: Rockwell Kent, op.cit., p. 27

p. 73 *a people kind through necessity:* ibid, p. 31

Chapter 8: Northern Light

p. 83 *Each minute of light experienced feels*: This evocation of Arctic summer sunlight is from Barry Lopez, *About this Life*, Chapter 7, "In a Country of Light, Among Animals," New York, Alfred A. Knopf, 1998, p. 122.

Chapter 9: The Passing of Whales

p. 93 *There are scores of them about us now*: Frederick Brueckner's fictional description of curious pilot whales, so similar to Tim Severin's description of actual whales penned some ten years earlier, appears in Frederick Brueckner, *Brendan (A Novel)*, San Francisco, Harper & Row, 1988, p. 111.

p. 94 *Brendan named the monster Jasconius*: The tale of Jasconius is retold here as it appears in Tim Severin's translation of the *Navigatio*, op. cit., Appendix I, p 246.

p. 94 *Looking down into the water, we could see:* ibid, p. 131

p. 96 *Long-finned pilot whales are a deep water species*: For a good description of the physical and behavioral characteristics of *Globicephala melaena*, see Wikipedia, "Pilot Whales," pp. 1–3, at *http://en..wikipedia.org/wiki/Pilot_Whale*. For a brief but accurate accounting of shore-based whale hunting in the Faroe Islands, see Wikipedia, "Whaling in the Faroe Islands," pp. 1–7, at *http://en.wikipedia.org/wiki/Whaling_in_the_Faroe_Islands*.

p. 104 *Indeed these "strong young men"*: For the whalers' hunt song, see Wikipedia, "Whaling in the Faroe Islands," loc. cit., p. 4.

Chapter 10: Footprints

p. 105 *Rule One: Carry out everything*: David Brower, ed., *The Sierra Club Wilderness Handbook*, New York, Ballantine Books, 1967.

p. 109 *For a few brief moments*: Details in this and following paragraphs about the disappearance and/or extinction of certain species of birds, fish, seals, bear, walrus, and other indigenous animal populations of coastal Labrador are from Farley Mowat, *Sea of Slaughter*, Boston, Atlantic Monthly Press, 1984.

p. 110 *Meanwhile the local population of humpback whales*: For a more detailed discussion of the near-annihilation and partial recovery of humpback whale populations along the Labrador and Newfoundland coasts, see Myron Arms, *Servants of the Fish*, Hinesburg VT, Upper Access, 2004, pp. 109–114.

p. 111 *As regards human disease*: For a review of the various human diseases carried into the Canadian Arctic by western visitors, including the poisoning of Inuit mothers' breast milk by industrial PCBs introduced into the food chain, see Anita Gordon and David Suzuki, *It's a Matter of Survival*, Cambridge (MA), Harvard University Press, 1991.

Chapter 11: A Gaunt Waste in Thule

p. 115 *The new Vale of Tempe may be a gaunt waste in Thule*: Thomas Hardy, *The Return of the Native*, New York, Harper & Brothers Publishers, 1922, pp. 5–6.

Chapter 12: Messages in the Earth

p. 127 *A brown land dark against the evening sky*: Rockwell Kent, *N by E*, New York, New York Literary Guild, 1930, p. 47.

pp. 131–36 *The story begins during the Proterozoic era*: The geological chronology of events in eastern Labrador that began two billion years BP and concluded with the arrival of the first human hunter-gatherers, about 9000 years BP, is derived almost entirely from Steven M. Stanley, *Earth and Life through Time*, Second Edition, New York, W.H. Freeman and Company, 1989, pp. 233–609. Certain details for local tectonic events in the "proto-Labrador" have been extrapolated from descriptions of larger Earth-wide events. Others have been filled in by means of personal communications with Steven Stanley. In a chronology as extensive as this one, however, not all researchers will agree on every detail. Although every effort has been made to compile the story as it is currently understood by a majority of scientific observers, the author accepts full responsibility for any errors or omissions that may be discerned, now or in the future, in the geological "story" as it appears in these pages.

Chapter 13: Ancestors

p. 137 *Environmental damage, climate change, rapid population growth*: Jared Diamond, *Collapse: How Societies Choose to Fail or Succeed*, New York, Viking, 2005. Lines quoted are from the dust jacket liner notes.

p. 137 *Each time the ground was broken*: Details in this and following pages about the discovery and excavation of the Maritime Archaic cemetery at Port au Choix, Newfoundland, are derived from several sources. First among these are interpretive materials displayed at the excavation site during the author's first visit there in July, 1988. Additional details appear in James Tuck, *Ancient Peoples of Port au Choix, Newfoundland*, Institute of Social and Economic Research, Social and Economic Studies No. 17, Memorial University of Newfoundland, 1976. A dramatization of Tuck's discoveries also appears in a documentary film, "Search for the Lost Red Paint People," conceived and written in part by William Fitzhugh and produced by PBS for its NOVA series.

p. 143 *Climate indicators in the far north are difficult to tease*: Details contained in these and following conversations with William Fitzhugh came by way of personal communications with the author during an interview conducted in Fitzhugh's offices at the Smithsonian Institution, Washington D.C., on February 18, 1994.

p. 144 *Summer after summer... these expeditions pushed ever farther north*: A detailed accounting of Fitzhugh's expeditions onto the Labrador coast can be found in William Fitzhugh, "Maritime Archaic Cultures of the Central and Northern Labrador Coast," *Arctic Anthropology*, Vol. XV, no. 2, 1978, pp. 61–95

p. 145 *All we can assume is that this new Eskimo population*: ibid, p. 91.

p. 147 *Likewise out at the newly discovered habitation site*: For an accounting of the search for and discovery of the Port au Choix habitation site, see M.A.P. Renouf and Trevor Bell, "Searching for the Maritime Archaic Indian Habitation Site at Port au Choix, Newfoundland: An Integrated Approach Using Archeology, Geomorphology, and Sea Level History," published by the Provincial Archeology Office, Department of Tourism, Culture, and Recreation, Government of Newfoundland. This report may be found online at *www.nfmuseum.com/9717Re.htm*.

Chapter 14: Milk Sea

p. 149 *Such high densities*: The description in the epigraph of coccolithophorae blooms in the surface layers of the open ocean is from King, Michael D. et.al. editors, *Our Changing Planet: The View from Space*, New York, Cambridge University Press, 2007. William Balch's contribution to this vast photographic compendium of changing Earth systems is a chapter entitled "Coccolithophores and the 'Sea of Milk,'" pp. 181–183. The quoted lines are found on p. 182.

p. 151 *The CZCS was launched:* Details about the duration and accomplishments of the CZCS can be found in Christopher Brown, "Global Distribution of Coccolithophore Blooms," *Oceanography*, v. 8, n. 2, 1995, pp. 59–60.

p. 152 *Among the researchers:* For details in this paragraph about data from the CZCS and the contributions by Patrick Holligan, see Balch, op. cit, p 183; also see Holligan et.al., "Satellite and ship studies of coccolithophorae production along a continental shelf edge, *Nature*, v. 304, 1983, pp. 339–342.

p. 152 *The story of the coccolithophorae blooms:* The discussion in this and following paragraphs of shell-building organisms and the chemistry that supports them is derived from two sources: Elizabeth Kolbert, "The Darkening Sea," *New Yorker*, November 20, 2006, pp. 66–75; and Ulf Riebesell, "Acid Oceans," *Our Planet*, December 2007, pp. 10–11.

p. 152–154 *The conditions known by the popular term:* Information on this and the following two pages about the distribution and life cycle of the coccolithophorae algae are from personal communications with Catherine Kilpatrick during a telephone interview conducted by the author on April 14, 1996.

p. 154 *The most comprehensive global biological record:* The quoted material in this and the following paragraph are from NASA *Earth Observatory*, news release 01–57, "First Chapter of Earth's 'Biological Record' Documented from Space,", March 29, 2001, pp. 1–2.

p. 154 *During the first three years of this mission:* Details in this and following paragraphs about the SeaWiFS mission are from lecture notes compiled by the author at a scientific colloquium presented by Sea-WiFS mission manager Charles McClain at NASA Goddard Space Flight Center, Greenbelt, Maryland, on April 17, 1998.

p. 155 *With three years of observations:* Remarks by Michael Behrenfeld quoted here are from NASA *Earth Observatory*, loc. cit., p. 2.

p. 155 *The prospect of linking the data:* Details about calibration for Sea-WiFS and later earth observing missions are from the Charles McClain lecture, as cited above.

p. 155 *In May 2002, with the launch:* The AQUA project's version of the MODIS was actually the second such instrument to be launched by NASA. The first MODIS was launched in December 1999 aboard the TERRA satellite. According to AQUA project manager Claire Parkinson, however, it is the AQUA MODIS that has been the more useful of the two for making ocean color measurements.

p. 157 *Oceanographer Ulf Reibesell explains*: Riebesell, op.cit. p. 11.

p. 157 *a significant decrease in the water's pH*: ibid p. 10.

p. 157 *The results of the experiment*: Ulf Reibesell et.al., "Reduced Calcification of marine plankton in response to increased atmospheric CO2," *Letters to Nature* 407, September 21, 2000, pp. 364–367.

p. 157 *There is a high risk that many*: Comments quoted early in this paragraph about the "loss of marine biodiversity" are from Reibesell, "Acid Oceans," p. 11.

p. 157 *In a more pessimistic moment*: Reibesell's concluding comment about the "rise of slime" is from his interview with Elizabeth Kolbert, op.cit., p. 75

Chapter 15: A Dune Adrift in the Atlantic

p. 159 *It's just sand… and it moves as though it were*: Marq de Villiers and Sheila Hirtle, *Sable Island: The Strange Origins and Curious History of a Dune Adrift in the Atlantic*, New York, Walker & Company, 2004, p. 3.

p. 159 *Instead, as core samples from a failed exploratory well*: Factual details about the geological history of Sable Island are from deVilliers and Hirtle, op. cit., Chapter Two: The Glacial Origins, pp. 23–44.

p. 160 *I learned about the more than 500 documented shipwrecks*: ibid, Chapter Four: The Serial Shipwrecks, pp. 52–57.

p. 163 *I came here in the 1970s to take part*: For background about the early years of the dune restoration program and other activities of Zoe Lucas on Sable Island, see Bruce Armstrong, *Sable Island*, New York, Doubleday & Company, 1981, pp. 185–189.

Chapter 16: The Beneficent Gene

p. 167 *I hammered on the gate, making the iron chain rattle*: Rory Stewart, *The Places In Between*, New York and London, Harcourt, Inc., p. 270.

p. 170 *I was alone… walking in very remote areas*: ibid, Dedication, p. iii.

p. 171 *"It is mid-winter," one of the officials told him*: ibid, p. 3.

p. 171 *I was passed like a parcel down the line*: ibid, p. 207.

p. 171 *For McWethy, the reasons for making*: Details in this and following paragraphs about David McWethy's journey across the northern United States in a horse-drawn gypsy wagon are from numerous personal conversations between the author and McWethy as well as from McWethy's diaries of the journey, privately published in 2005 under the title *Blessings of the Open Road*, and made available to the author as a gift.

p. 172 *the ancient Egyptian "Story of Sinhue"*: For a more detailed discussion of this ancient paradigm for Bedouin hospitality, see A.F. Rainey, "The World of Sinuhe," *Israel Oriental Studies II*, ed. Richard Walzer, Tel Aviv, 1972, p. 372.

p. 173 *When he [Abraham] saw them:* Genesis 18: 1–8.

p. 173 *Do not neglect to show hospitality*: Hebrews 13:2.

p. 173 *Elsewhere in the ancient Mediterranean world*: For a discussion of Homeric hospitality, see S. Reece, *The Stranger's Welcome*, "Oral Theory and the Aesthetics of the Homeric Hospitality Scene," Ann Arbor, 1933. pp. 3ff.

p. 173 *For a kindness that may seem small to us*: Odyssey, Book VI, lines 230–232.

p. 174 *In the Koran, sustenance is considered*: For traditions of hospitality in the Islamic world see M. Schulman and A. Barkouki-Winter, *The Extra Mile*, especially Part 3: "Duty and Superfluity," as published online by Santa Clara University at *www.scu.edu/ethics/publications/iie/v11n1/hospitality.html*.

p. 174 *In the Torah, the faithful*: For the tradition of *tzedakah* (charity) in early Jewish life, see Frank Loewenberg, *From Charity to Social Justice: The Emergence of Communal Institutions for the Support of the Poor in Ancient Judaism*, Dimensions, 2001.

p. 174 *the ancient Celtic tradition of hospitality*: For a discussion of the historical and legal mandates for *oigidecht* (hospitality) and the *bruideans* in ancient Ireland, see Peter Berresford-Ellis, *A Brief History of the Celts*, London, Constable & Robinson, Ltd., 1998, 2003. Also see Fergus Kelly, *A Guide to Early Irish Law*, Dublin Institute for Advanced Studies, 1988.

p. 174 *In Poland, the traditional Christmas eve supper*: A description of leaving an empty place set for a stranger or "uninvited guest" during the Polish *wigilia* can be found on-line at *http://en.wikipedia.org/wiki/Wigilia*.

Bibliography and Further Reading

Following is a listing of the most important sources used in writing these essays, as well as a selection of other titles that may be of interest to readers wishing to investigate further some of the ideas they have encountered in these pages.

Armstrong, Bruce, *Sable Island: Nova Scotia's Mysterious Island of Sand*, New York, Doubleday Company Inc., 1981.

Berry, Thomas, *The Dream of Earth*, San Francisco, Sierra Club Books, 1990.

Buechner, Frederick, *Brendan (A Novel)*, San Francisco, Harper & Row, 1988.

Carson Rachael, *The Sea Around Us*, New York, Oxford University Press (Special Edition), 1989.

Conway, Jill Ker, *The Road from Coorain: Recollections of a Harsh and Beautiful Journey into Adulthood*, New York, Alfred A. Knopf, 1989.

DeVilliers, Marq and Hurtle, Sheila, *Sable Island: The Strange Origins and Curious History of a Dune Adrift in the Atlantic*, New York, Walker & Company, 2004.

Diamond, Jared, *Collapse: How Societies Choose to Fail or Succeed*, New York, Viking, 2005.

Dillard, Annie, *Pilgrim at Tinker Creek*, New York, Harper & Row, 1974.

Elder, John, *Reading the Mountains of Home*, Cambridge (MA), Harvard University Press, 1998.

Fagan, Brian, *The Little Ice Age: How Climate Made History 1300–1850*, New York, Basic Books, 2000.

Hardy, Thomas, *Return of the Native*, New York, Harper & Brothers Publishers, 1922.

Holdridge, Desmond, *Northern Light*, New York, The Viking Press, 1939.

Kent, Rockwell, *N by E*, New York, New York Literary Guild, 1930.

King, Michael D. et. al., eds., *Our Changing Planet: The View from Space*, New York, Cambridge University Press, 2007.

Kolbert, Elizabeth, "The Darkening Sea: What Carbon Emissions are Doing to the Ocean," *New Yorker*, November 20, 2006, pp. 66–75.

Lopez, Barry, *About this Life: Journeys on the Threshold of Memory*, New York, Alfred A. Knopf, 1998.

_____, *Arctic Dreams: Imagination and Desire in a Northern Landscape*, New York, Charles Scribners' Sons, 1986.

_____, *Crossing Open Ground*, New York, Vintage Books, 1989.

McPhee, John, *Annals of the Former World*, New York, Farrar, Straus and Giroux, 1981, 1983, 1993, 1998.

Mowat, Farley, *Sea of Slaughter*, Boston, Atlantic Monthly Press, 1984.

_____, *The New Founde Land*, Toronto, McClelland and Stewart, 1989.

Norris, Kathleen, *Dakota: A Spiritual Geography*, Boston, Houghton Mifflin Company, 1993.

Riebesell, Ulf, "Acid Oceans," in *Our Planet*, December 2007, pp. 10–11.

Severin, Tim, *The Brendan Voyage*, London, Hutchinson & Co., 1978.

Stanley, Steven M., *Children of the Ice Age: How a Global Catastrophe Allowed Humans to Evolve*, New York, W. H. Freeman and Company, 1996.

_____, *Earth and Life Through Time*, Second Edition, New York, W. H. Freeman and Company, 1989.

Steward, Rory, *The Places In Between*, New York and London, Harcourt, Inc., 2004.

Personal Acknowledgments

Above all, I would like to thank the ninety-two men and women who have served as crew of *Brendan's Isle* on her various journeys into the northern seas over the past twenty-five years. Some have sailed only small parts of the way. Others have been there for major portions of the journey: Andrew Nyhart, Kell Achenbach, John Griffiths, Mikey Auth, Amanda Lake, Steve Arms, Andrew Wilson, Tim Clark, Beth Perry, Scott Harris, Kirk Fitzsimmons, Pete Johanssen, Mike ("Bluey") Brown, Liz Bogel, Tommy Knowles, Jan Scott, Terry and Maggie Duckett, Cherie Patricelli, Katie Harris, Ben Gray, Bill Alberts, Joe Haran, Nate Parcells, John and Fran Miller, and of course Kay, who has been there, both in person and in spirit, for every mile *Brendan's Isle* has sailed.

Special thanks also go to the many individuals who have read and helped to shape these essays at various stages of their evolution, and especially to Steve Stanley, Clair Parkinson, John Miller, Joe Haran, Bill Alberts, Bernadette Bernon, Elaine Lembo, John Burnham.

The photographs that appear throughout the book are the work not only of the author but also of talented crew members who have documented their own portions of the journey: Kay Arms, Jess Rice, Mikey Auth, Mike Brown, Will Barker, Joe Haran, Nate Parcells, Silver Donald Cameron.

I owe a huge debt of gratitude to my publisher/editor Steve Carlson for his careful and incisive editing of my prose as well as for his patience, encouragement, and support in moving this project from a rough manuscript to a finished book.

And finally, and most emphatically, I wish to thank graphic designer Kitty Werner for her truly inspired work on the cover design, the maps, the photographs, and the overall page design for the book. Kitty's critical eye and creative mind are everywhere evident in the look and feel of the beautiful volume she produced.

About the Author

Ever since their marriage in the early 1960s, Myron (Mike) Arms and his wife Kay have been dreaming up ways to go sailing together. Beginning with their week-long honeymoon on a borrowed sailboat, they have followed the dream for more than 40 years.

In the early 1970s they stepped out of busy lives ashore to home-school their three children during a nine-month voyage from the northeastern United States to the high islands of the Caribbean. Later, with a new and stronger boat, they turned their sights northward, eventually logging more than 140,000 blue-water miles, including two high-latitude crossings of the North Atlantic, a voyage to western Greenland, a circumnavigation of Iceland, four summers exploring Scandinavia and the British Isles, and numerous voyages to the Atlantic coasts of Newfoundland and Labrador.

Arms is the author of four previous books, including *Servants of the Fish* and *Boston Globe* bestseller *Riddle of the Ice*. He has published more than fifty feature articles in *Cruising World, Sail, Blue Water Sailing*, and many other sailing and adventure magazines. Readers may sample these and other writings and may follow the Arms' future sailing adventures on the Web at www. myronarms.com.

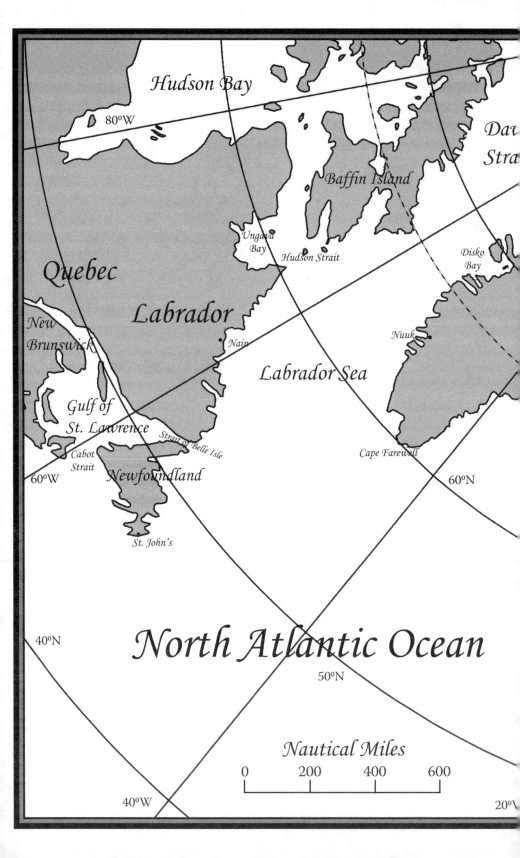